Corpse

CORPSE

Nature, Forensics, and the Struggle to Pinpoint Time of Death

JESSICA SNYDER SACHS

PERSEUS
PUBLISHING

Cataloging-in-Publication Data is available from the Library of Congress
ISBN 0–7382-0336-X

Perseus Publishing is a member of the Perseus Books Group.
Find us on the World Wide Web at http://www.perseuspublishing.com.
Perseus Publishing books are available at special discounts for bulk purchases in the United States by corporations, institutions, and other organizations. For more information, please contact the Special Markets Department at the Perseus Books Group, 11 Cambridge Center, Cambridge, MA 02142, or call (617) 252-5298.

Set in 11-point Fairfield Light by Perseus Publishing Services

First printing, September 2001

1 2 3 4 5 6 7 8 9 10—03 02 01

This book is dedicated to the memory of
Greg Pearson (1932–1989), *first-class journalist,*
uncompromising teacher, and beloved mentor,
and
Lamar Meek (1944–2000),
forever a gentleman and a scholar.

Contents

Acknowledgments

THE NAMES OF my many coauthors could never fit on a book jacket, let alone a single page. I can only begin to thank the scores of scientists, librarians, and media information specialists who graciously submitted themselves to my endless questions and requests for obscure historical and scientific information. Special thanks go to the pathologists, anthropologists, entomologists, and botanists who have reviewed my accounts of their work. Any inaccuracies that have managed to survive their vetting remain utterly my own. For reasons unknown but greatly appreciated, several of these researchers have taken an extraordinary interest in my work, with patient tutoring and feedback that goes far beyond their too-brief mention in the text. They include most especially Germany's peerless "Herr Maggot," Mark Benecke, Virginia Commonwealth University's Jason Byrd, Rob Hall of the University of Missouri, "Wild Man" Neal Haskell of Saint Joseph's College, and retired USDA entomologist Jerry Payne.

Thanks also go out to Joyce Adkins, widow of the late forensic entomologist Ted Adkins, for trusting me with her husband's papers and David Faulkner, of the San Diego Natural History Museum, for trusting me with his own; Tracy Cyr for our e-mail discussions of insect minutiae; forensic anthropologists Bill Bass, William Haglund, and Clyde Snow for the delightful ways they spin their tales; forensic pathologists Werner Spitz, Jay Dix, and Stephen Cina for revealing the imperfections as well as the strengths of their science; and Regius Professor Peter Vanezis of the University of Glasgow for sharing the nineteenth-century history of his great institution. Invaluable research assistance came my way from the public relations departments of the University of Illinois at Chicago, Washington State University at Pullman, and Clemson University, as well as Professor Hassan A. Babaei of Georgia State University, Sarah Goodwin of Emory University, and the reference service of the Atlanta-Fulton Public Libraries.

On the literary side of this endeavor, my thanks go out to my wonderful agent, Regula Noetzli, who risked alienating her many publishing contacts by raising this book's grisly subject matter over lunch at many fine Manhattan restaurants, and to my amazing editor at Perseus, Amanda Cook, who saw past the disgust factor to grasp the potential beauty of this book, came up with the perfect name for it, and patiently forged an author out of a journalist. I also wish to thank former *Discover/Science Digest* colleague Jeffrey Kluger (*Apollo 13, Journey Beyond Selene*) for helping me navigate the ethical minefield of re-creating history from the fading memories of its participants and their survivors.

Finally and most important, my intermingled love and gratitude go to my parents for their encouragement, to my daughter Eva for her never-ending delight in the "little grubbies," and to my soul mate Gary, for tolerating the parade of odious photos across my desk as well as my newfound interest in flies on dead squirrels and picnic plates.

CORPSE

Prologue

On the evening of November 8, 1983, friends and relatives of Susan Hendricks were reeling from the news that police had found the soft-spoken woman and her three young children hacked to death in their beds. What could they say to David, the affectionate husband and doting father, who had just arrived home from an out-of-town business trip to find squad cars swarming around his home in the suburbs of Bloomington, Illinois? Their shock was only compounded when police let it be known that David Hendricks was their prime suspect.

"There's just no way David could kill those children or Susie," the children's stunned grandmother told a reporter as she watched two hearses pull out of the family driveway, orange body bags visible through the long, tinted windows. "He loves them. They're a perfect family."

In the sensational murder trial that followed, relatives on both sides of the Hendricks family as well as members of their close-knit Christian fellowship rallied behind the accused. They described a considerate husband and charitable businessman who had recently thrilled his wife with a romantic tenth wedding anniversary trip to England and who funneled tens of thousands of dollars to the needy from the sales of his patented orthopedic braces. Neighbors described how, the evening before the murders, Hendricks had taken his children to see his nine-year-old, oldest daughter Becky's prize-winning artwork hanging at a local mall, then to Chuck E. Cheese Pizza Time Theater, where food was secondary to the raucous indoor playground.

After putting his children to sleep that night, Hendricks testified, he had waited for his wife to come home from a baby shower, kissed her good-bye around 11 P.M., then drove through the night to Wisconsin, where he'd spent the next day making sales calls. The real killer or killers, Hendricks's defense attorneys claimed, had slipped into the house through an unlocked door sometime after midnight and slaughtered Susan and the three children in a burglary rampage. Still on the loose, the cold-blooded killers presented a clear and alarming danger to unsuspecting families everywhere.

Police and prosecutors painted a very different picture. Hendricks, they argued, had carefully planned the murders, killed his family as they slept, ransacked his own home to make it look like burglary, then coolly drove away from the carnage. That no fingerprints were found on the murder weapons—an ax from the garage and a butcher knife from the kitchen—only supported their contention that this was a premeditated murder. Hendricks's motive: freedom from the constricting bonds of marriage without sacrificing his high standing in a church group that forbid divorce except in cases of adultery. As evidence of Hendricks's double life, the police found several women whom he had hired to model his back braces for advertising brochures—women who said Hendricks had touched them in ways that made them feel uncomfortable. As further evidence of Hendricks's secret desires, the prosecution focused attention on his recent weight loss, new hairstyle, and sudden interest in fashionable clothes. Police also described the creepy calm with which Hendricks had received the news of his family's slaughter.

Which persona, which scenario was the jury to believe? In the end, their decision had to turn on the "when" of the horrific murders. Were wife and children sleeping peacefully when David Hendricks backed out of his driveway that night, or were they already lying dead atop their blood-soaked beds?

To answer this question, a parade of North America's most renowned medical examiners filed through the Winnebago County Courthouse in October 1984, each of them a veteran of more than a thousand forensic autopsies. The prosecution's first expert witness, Dr. Michael Baden of the New York City medical examiner's office, spent nine hours on the witness stand explaining how a forensic pathologist tries to determine time of death based on findings at autopsy and how he personally came to the conclusion that the Hendricks children were dead by 9 P.M. the night of the murders. Given the interval of more than a day between the murders and the subsequent autopsies, the postmortem processes of body cooling, stiffening, and blood pooling proved no help, he admitted. Fortunately, the local pathologist had carefully removed, measured, and stored the contents of each victim's stomach.

At this point in the trial, a prosecutor brought four large Styrofoam cups to the witness stand. Baden opened each in turn, held them to his nose, and identified them as the gastric contents of Susan Hendricks and her three children. He described how the quantity and quality of the foodstuffs in the children's stomachs—fragments of mushrooms, olives, and onions—indicated they had died approximately two hours after finishing their 7 P.M. pizza. Three more prosecution witnesses, including the chief medical examiner of Dade County, Florida, and the chief toxicologist for the Illinois Department of Public Health, supported Baden's opinion.

But the battle of medical experts had just begun. The defense called to the stand four equally renowned forensic pathologists who testified that time of death could never be pinpointed by something as variable as digestion. Although a typical meal takes two to four hours to pass out of the stomach, a hundred different factors could speed or delay the process, they argued. Factors such as the excitement children experience during a special night on the town? asked a defense attorney. Factors such as gulping food in large bites, as children do when they're in a rush to get back to their play? Factors such as strenuous physical activity, say, jumping on an air-pillow trampoline, climbing rope ladders, and "swimming" through a vat of plastic balls?

Earlier, the prosecution's own medical witnesses had discounted the possibility that such "normal" activities would significantly impact the children's digestion. "If it did, my seven children and eleven grandchildren would be suffering from indigestion all the time," one had quipped. By contrast, the witnesses for the defense argued that the medical literature supported their opinion that such things can and do slow digestion, though not in a predictable way. One cited the example of an eleven-year-old murder victim known to be alive more than four hours after he ate a pizza lunch at school, a boy whose stomach contents on autopsy were more than double the amount of the stomach contents of the three Hendricks children combined.

What was the jury expected to make of such conflicting testimony? A parade of top pathologists had come before them, equally divided as to the meaning, even the value, of the only medical evidence indicating when Susan Hendricks and her children met their deaths.

* * *

TIME OF DEATH. Throughout the long annals of true crime lore, countless murder convictions and acquittals have come down to this: When did the killer strike? When did the victims breathe their last? In the absence of credible witnesses, the lack of an easy answer has bedeviled our criminal justice system since its inception.

Admittedly, before the eighteenth century, few Western courts bothered to quibble about such abstract concepts as proof of guilt or even credible evidence of it. Portuguese traders returning from China in the sixteenth century marveled at the lengths that Chinese magistrates went to avoid condemning a person to death. By contrast, torture was standard procedure for eliciting a murder confession throughout Europe, assuming a hatchet man or lynch mob hadn't dispatched the accused before he made it behind bars.

Our modern criminal justice system, which sets the deliberately high hurdle of "reasonable doubt" before conviction, largely took form in the aftermath of the French Revolution with its *Declaration of the Rights of Man and of the Citizen* (1789). As applied to crime and punishment, these rights established the principle of "innocent until proven guilty" in an open court of law. Perhaps nowhere was this change so profoundly felt as in the investigation of homicide. No longer was it enough to drag a degenerate-looking scoundrel into court. In the absence of witnesses, police and prosecutors had to justify arrest and trial by showing that a suspect had not only the motive and means to kill, but also the opportunity to do so. In other words, they had to offer judge or jury a plausible story placing victim and accused together at the time the fatal "blow" was delivered.

Consequently, murder investigators found themselves desperate for clues as to time of death, and not just for evidence of guilt at trial. Knowing when a victim died could speed the earliest stages of an inquest by ruling out suspects with confirmed alibis and focusing scrutiny on those who did not. The *postmortem interval,* or time since death, proved even more critical in cases where a corpse turned up decomposed beyond recognition. Even an approximate time of death gave investigators a framework in which to connect the remains to a suspicious disappearance.

Yet for all its importance, determining time of death has defied the detective's magnifying glass and the pathologist's scalpel for over 2,000 years. Even today, despite crime labs crammed with high-tech equipment for DNA analysis, toxicology, serology, and the detection of rarefied chemical vapors, we remain nearly as blind as the ancient Greeks with their belief in maggots sprouting fully formed and spontaneous from the flesh of the newly dead.

Nonetheless, it still startles most people to learn that a prudent medical examiner can rarely, if ever, accurately measure the interval between death and a body's discovery. Murder mysteries and crime shows certainly give the subject short shrift. Even TV's intrepid *Quincy, M.E.* somehow avoided the question, spending episode after episode cleverly teasing out *cause* of death for his storied corpses.

In real life, the challenge tends to be just the opposite. Cause of death is usually more than obvious to every police officer responding to the scene of murder. Knives leave gaping wounds, bullets blackened pits, and clubs a crumpled skull. Even poisons leave traces—child's play for a coroner with access to the most basic toxicology screens. But if the body contains a trusty chronometer—some biochemical display set blinking when the power goes off—it has yet to be found.

None of this is to say that the field of forensic pathology has ever relented in its quest for the perfect postmortem clock. By the mid-nineteenth century, European pathologists had described the classic trio of stopwatches still used today: rigor mortis, algor mortis, and livor mortis. The best known of the three, rigor describes the gradual muscle stiffening that grips a body in the hours after death and then melts away just as gradually. Algor represents the slow cooling of a warm-blooded corpse as it equilibrates with the temperature of its surroundings. The red-purple stain of livor, or lividity, documents the gradual settling and pooling of blood that begins the moment blood pressure plummets to zero.

But from their first use, the pathologist's three standard timepieces have proven unreliable, plagued as they are by death's infinite variations. Age, body size, health, manner of death, ambient temperature, air movement, even something as seemingly ineffable as the agony of a victim's final moments has been found to skew the body's postmortem changes beyond predictability. Even if their rates were predictable, the pathologist's clocks unwind all too soon. Within twenty-four to forty-eight hours after death, lividity reaches its peak, the body reaches room temperature, and rigor disappears. Two days after death, pathology's three timepieces—already only marginally helpful—become useless.

Twentieth-century pathologists added stomach contents to the factors that might suggest time of death. Unlike the three clocks set in motion at death, digestion slams to a halt. In theory at least, a medical examiner who knows the approximate time and quantity of the victim's last meal should be able to extrapolate time of death based on the rate that food might be expected to pass out of the stomach and into the intestines. In reality, the vagaries of stomach emptying in the soon-to-be-dead have proven even more problematic than the postmortem markers of rigor, livor, and algor.

Throughout the second half of the twentieth century, pathologists and toxicologists have continued to fill forensic science journals with reports on potential new indicators of time since death. Some of the most promising have included a gradual rise in potassium levels in the eye's jellylike vitreous humor and the waning ability of the cadaver's muscles to respond to mild electric shock. Yet every such stopwatch has eventually proved inadequate when applied to actual murder, in all its controlled-experiment-defying depravity.

Still, the myth of the medical expert's ability to nail down time of death has endured. No doubt this stems in part from the many pathologists who continue to offer more precision in court than their science can rightfully claim. That they do so is understandable enough, given the relentless pressure. "It's a question almost invariably asked by police officers, sometimes with touching faith in the accuracy of the estimate," wrote famed English pathologist Bernard Knight in the 1960s. "It's one of the most common questions I get," echoed Missouri medical examiner Jay Dix forty years later. "I have to tell them—it's impossible." Yet Dix—one of the nation's top pathologists and the author of the 1999 forensic atlas

Time of Death, Decomposition, and Identification—sees it done all the time. "I'm continually reviewing cases in which pathologists pinpoint death to within a few hours," he said. "Not that I've ever seen a case where it was appropriate."

Indeed, Quincy isn't the only medical examiner who can sound convincing on the witness stand. Among the many questionable verdicts based on dubious time-of-death testimony stands a murder conviction that still ranks as Canada's most controversial—one involving a fourteen-year-old Ontario boy sentenced to death for the rape and strangling of a twelve-year-old classmate. The conviction hinged largely on a medical examiner's opinion that the victim had died during a half-hour interval when the two were together, a determination based solely on the food removed from the girl's stomach at autopsy. Today, most prudent pathologists scoff at the naïveté, if not misconduct, of anyone claiming to pinpoint time of death so precisely based on stomach contents. (Even Baden gives a fudge factor of plus or minus two hours in his determinations.) Yet many forensic doctors make comparable judgments based on similarly questionable postmortem clocks.

"If a pathologist says the death occurred at such and such a time, I say get out the handcuffs, he was there," contends medical examiner Stephen Cina of the Armed Forces Institute of Pathology. Given the sad state of the art, one could argue that extrapolating time of death in the absence of witnesses should be altogether banned from the courtroom.

But there's a quiet revolution afoot in forensics. Surprisingly, it does not draw on twenty-first-century medical expertise, biochemical assays, or even computerized analysis for answers, but on the "softer" natural sciences. A new mod squad of entomologists, anthropologists, and botanists are joining forces to tackle the age-old problem of postmarking death. Already their field—*forensic ecology*, if you will—has made more progress in solving the death investigator's greatest enigma than have 2,000 years of criminology and autopsy.

Anthropologists were the first to cross over from the natural sciences to forensics. In America, the fateful jump came in the 1930s, when FBI agents setting up the bureau's first crime lab in Washington, D.C., discovered a whole nest of "bone detectives" in the red Gothic towers of the Smithsonian Institution, across the street. As the curators of one of the world's largest collections of human skeletons, the Smithsonian anthropologists were uniquely qualified to help the FBI distinguish human from animal remains. From the identification of bones as human, forensic anthropology quickly advanced to the identification of individuals, based on distinguishing bumps and bony scars left by past injuries and the wear and tear of daily toil (a milkmaid's worn elbow, a tailor's notched thumb, and a mailbag carrier's crooked spine).

But anthropologists quickly realized the near impossibility of naming the dead without some method, however crude, of matching their identity clues to missing person reports for a given span of time. The most experienced among them could sometimes come up with a reasonable estimate of time since death by "feel"—that admittedly nonscientific second sense based on a lifetime of processing decayed corpses and crumbling bones. But precious few ever attempted the monumental task of objectively studying the stages that mark a human body's passage back to dust. So far, the most valuable dating method to come out of their research belongs by all rights to another science.

In the 1980s, the field of forensic entomology burst on the scene as if out of nowhere when bug and bone scientists independently discovered the value of what may be nature's ultimate postmortem clock—the cadaver-feeding insect. Maggots, once routinely washed from the coroner's table with disgust, suddenly became the hot new thing in homicide investigation. Still, the extent of the bugs' testimony had yet to be fully fathomed.

As anthropologists and entomologists began teaming up in their forensic investigations, they naturally turned to a third specialty to make sense of the roots and vines winding through their death scenes: A delicate green tendril snaking through a sun-bleached skull. A tree growing down through a shallow grave in the woods. A flush of growth marking the outlines of an inexplicably fertile corner of an abandoned lot. Each became yet another promising measure of the seasons that follow "death most foul."

Mordre wol out (murder will out), Chaucer's fourteenth-century prioress declared. "But those who work in homicide investigation, forensic pathology, and criminal law know better," mocked former U.S. Attorney General Ramsey Clark in his 1973 preface to *Spitz and Fisher's Medicolegal Investigation of Death,* forensic pathology's enduring bible. "The true manner of death which may have been murder is not determined in tens of thousands of cases annually in our violent land," Ramsey continued. "The cost to the nation in truth, justice, health and safety is enormous."

Ironically, it may be a return to Chaucer's touching faith in the constancy of nature that gives forensics its long-sought-after Holy Grail. It may well be that nature trumps technology in producing death's infallible stopwatch.

Corpse is the story of this pursuit and journey—the birth of a trio of natural sciences that together address what might be called the forensic ecology of human remains. It is likewise an account of their long road to acceptance in the courtroom, their maturation, their recent triumphs, and their future challenges. Like the smell of death that clings to the nostrils for days on end, it is a story that piques and haunts.

1 THE BODY HANDLERS

The psychiatrist knows nothing and does nothing
The surgeon knows nothing and does everything
The pathologist knows everything . . .
but is always a day too late.
—TRADITIONAL MEDICAL MAXIM

THE TYPICAL AMERICAN goes into the ground injected with three to four gallons of preservatives. But a sizable segment of even our oversanitized culture will always escape quick processing. Prominent among this population: the abandoned and the murdered. In theory, their moldering bodies—slumped under bridges, forgotten in bed, or dumped along roadsides—retain the natural if repulsive clues that might disclose time of death. For reasons as sensible as sensory, police are quick to pass these unvarnished dead to the next in line of custody—the coroners and medical examiners whose job it is to coax secrets from a corpse.

Many trace the work of forensic pathology—the medicolegal investigation of death—to the ancient Greeks who, circa 380 B.C., began dissecting various animal carcasses and applying their findings—at times absurdly—to humans. Greeks physicians did perform the rare human dissection. But Hippocratic writings express a deep disdain for the business, even as they named it: *autopsy.* Despite its strange implications of self-examination, the term has resisted 2,000 years of scientific lobbying to replace it with the more logical *necropsy.*

The Egyptians, meanwhile, suffered no such Hippocratic qualms. Historic accounts of anatomy and pathology classes at the Museum of Alexandria in the third and fourth century B.C. describe not only autopsies but also the live dissection of criminals "for the study, even while they breathed, of those parts which Nature had before concealed." Among the Greek and Egyptian observations were descriptions of death's first known clocks: *rigor mortis,* or postmortem stiffening, and *algor mortis,* body cooling. Unlike their nineteenth-century counterparts, who would chart the stiffening and cooling with hour-by-hour muscle testing and body-core thermometers, the ancients were satisfied with the postmortem touch test that homicide detectives still use today:

Warm and not stiff: Not dead more than a couple hours
Warm and stiff: Dead between a couple hours and a half day
Cold and stiff: Dead between a half day and two days
Cold and not stiff: Dead more than two days.

The physiology behind the rigidity and cooling rate would elude understanding for two millennia. In the meantime, scientists and nonscientists alike clung to many postmortem myths supplied by the early Greeks and Egyptians. Among them, the belief that rigor could make a corpse sit up in bed or clench its fists in undead rage. (In reality, rigor causes muscles to stiffen, not actively contract.) The ancients also misinterpreted a legitimate artifact of rigor—that of hair standing on end—as a sign that whiskers and curls continued to grow after death. If that were true, they would have discovered a postmortem clock that could literally be measured with a ruler. Today, pathologists understand that a cadaver's hair at times *appears* longer because of the stiffening of the tiny muscles surrounding each hair follicle. The same phenomenon can produce a set of head-to-toe goosebumps that has *nothing* to do with what the victim saw at the moment of death.

History records the first known application of medical knowledge to death investigation in 44 B.C. Summoned to examine the body of Julius Caesar, the Roman physician Antistius announced that he knew which of the would-be emperor's twenty-three stab wounds had proved fatal. By clocking death to a particular blow, Antistius thwarted the plot by which the Roman senators had hoped to avoid any *one* of them standing trial for murder. In the end, history tells us, they all paid with their lives. But Antistius's historic death determination, however dubious it may have been, marked the beginning of the pathologist's role as expert witness to murder. In fact, it gave us the term *forensic,* Latin for "before the forum," which is where Antistius made his fateful declaration.

Meanwhile, half a world away, a nonmedical system of death investigation had taken root in China. Bamboo slips unearthed at sites dating to the Ch'in dynasty (221–207 B.C.) detail the procedures that civil servants were required to follow when summoned to examine a corpse found under suspicious circumstances. China's state-ordered death investigations grew in sophistication during the progressive Sung dynasty (A.D. 960 to 1279), culminating in the earliest known forensic handbook by Hsi Yuan Chi Lu, *The Washing Away of Wrongs,* in 1247.

In it, the death investigator Sung Tz'u describes how to ascertain time of death in each of the four seasons, by assessing the extent of a corpse's decomposition. He notes, for example, that during the cool months of spring, a visible reddening of the mouth, nose, and belly indicates a postmortem interval of two to three days. "After ten days, a foul liquid issues from the nose and ears," he continues, cautioning the homicide detective to take into consideration the build and previous health of the victim. "In fat and swollen people it is like this. Those long ill and emaciated will display these symptoms only after half a month."

In summer, the same sort of purging from the nose and mouth postmarked death three days past, with hair falling out on the fourth or fifth day. By contrast, freezing weather slowed decomposition to a crawl. "During the three winter months, when four of five days have passed, the flesh of the corpse will turn yellowish purple. After half a month, the symptoms [reddening] described above will appear." In general, Sung Tz'u instructed death investigators to multiply by five days the time required for the colors and smells of decomposition to appear in winter as compared to summer. Although the rubric ignored the many hidden factors that could speed or slow human decay, it accurately captured the pivotal role of bacterial growth, which accelerates in warm weather to hasten the breakdown of soft tissue.

Sung Tz'u refrained from attempting to pinpoint time of death to any window narrower than a day or two. His discretion appears all the more wise, given the long distances China's thirteenth-century death investigators had to travel on foot to cover their large districts. More often than not, their examinations involved bodies in advanced stages of decomposition. Yet even the crude postmortem markers of advanced decay could help sort a list of suspects by alibi and opportunity. More important, perhaps, a time-stamped trail of corpses gave the Sung constabulary some hope of tracking down the roving gangs of bandits that plagued thirteenth-century China, with its population of 100 million spread over a land area comparable in size to that of western Europe.

Yet Sung Tz'u and his students rendered their opinions without the benefit of any medical training. Indeed, physicians played little to no role in ancient Chinese homicide investigations, a situation that Asian legal scholar Brian McKnight attributes to their low social status. In this aspect, Chinese forensics had a clear counterpart in the nonmedical coroner system that appeared in England in the tenth century. McKnight even suggests a direct influence, with information about the Chinese system passed to England by way of the well-traveled courtiers of Sicily.

In any case, we know that in 925, King Athelstan appointed the English noble St. John of Beverly to be the first "keeper of the please of the crown," an unwieldy title soon truncated to "crowner" and eventually "coroner." Then as now, the legal basis for the coroner system was to "enquire where, *when,* and by what means, a person came to his death." Admittedly, the Old English coroner's need to postmark death had more to do with record keeping than criminalistics. Registering the death date for a person who died in his sleep required the coroner to ascertain whether that person expired before or after midnight. The official chronometer in such cases would have been the back of the coroner's hand held against the corpse to judge its warmth or lack of it. Occasionally, such determinations had material consequences, as when brothers or other joint heirs died in battle, with all inheritance passing to the family of the last to die.

Outside such time judgments, the ancient coroner's primary interest centered on signs of suicide, a crime against God and king that resulted in forfeiture of the victim's estate, a handy source of royal income that spelled destitution for survivors. Unfortunately, the dire consequence of this duty quickly led to the coroner system's corruption. By Shakespearean times, the coroner's susceptibility to bribes had become a well-known joke, as illustrated in act 5, scene 1, of *Hamlet,* in which two grave diggers laugh at the crowner's pronouncement that Ophelia's suicidal drowning had been accidental—owing to the water coming to her, not her to the water. Historical records suggest that Shakespeare lifted this twisted bit of logic from an actual coroner's ruling of his day.

England's corrupt coroner system would prevail throughout the British Empire, including present and former colonies, for centuries—greatly slowing medical advances in death investigation. Expanding the scientific roadblock was the Council of Tour's twelfth-century condemnation of autopsy as "an abomination against God."

Yet in the face of all this, the first quasi-medical examiner system appeared in 1532, when Emperor Charles V decreed that medical testimony be part of all trials involving "homicide, infanticide, abortion, or poisoning" throughout his Holy Roman Empire of central and southern Europe. Their new role in legal proceedings gave Europe's early anatomists new legitimacy. It also gave them reason to look differently at the cadavers they had been secretly dissecting to better understand the living. Could the cold flesh also reveal the manner and timing of its death?

By the end of the century, the French physician Ambroise Paré had recorded Europe's first official criminal autopsies, describing among other things the lungs of deliberately smothered children (fluid-filled and speckled with blood). Over the next generation, the new field of forensic medicine came into being with the publication of a succession of studies, culminating in Paolo Zacchia's *Quastiones Medico-Legales.*

Importantly, Paré, Zacchia, and their contemporaries added to rigor and algor a third postmortem timepiece: lividity, or livor mortis, from the French *liviere,* "to turn blue." What they had documented was the gradual deoxygenation and gravitational settling of the blood that begins as soon as lungs and heart cease their motions. Importantly, lividity—with its clear and predictable procession of hues from pink through purple to blackish-blue—presented a stopwatch visibly set in motion within minutes, not just hours, of death.

The color progression of lividity begins with the proverbial pallor of death in an already light-skinned person, as blood begins to drain out of the upper surfaces of the body. As soon as fifteen to twenty minutes after death, an experienced observer can see the first diffuse blotches take form on the underside of the body. The seepage likewise becomes visible in dead-end crannies such as earlobes and skin folds. Within an hour or two, the telltale discoloration becomes obvious to even the untrained eye. The pink "slap" of early livor gradually darkens to a dull, bruiselike red before progressing through shades of purple and blue as oxygen gradually disappears from the blood.

However, the lividity is not yet "fixed," or permanent. Press your thumb against an area of livor in the first hours after death, and it will blanch. Similarly, should you move the body during this period, the blood-settling patterns will shift, though perhaps not completely, for livor's fixation is not all or nothing, but gradual. A body dead in a kitchen chair at 5 P.M., then undressed and tucked into bed at 8 P.M. may retain the faintly blanched impressions of contact points between the body and unyielding surfaces such as the back of that chair or a tight waistband.

By ten hours past death, lividity's stain has become fully fixed. The body has now cooled to the point where the fatty lining of the blood vessels congeals, pinching shut the tiny capillaries near the body surface. The dark stain of blood seepage can no longer escape inward when pressed, nor will it resettle, even partially, when the body is shifted. Moving a body once livor has fully set leaves behind a stark and permanent imprint of death's original position. Oftentimes, the detail can reveal the very texture of the surface on which the victim dropped—be it the stippled inscription of a gritty path, the weave of a carpet, or the design of kitchen linoleum. (Indeed, the pale imprint of the toilet seat across the buttocks of heart-attack victims helped twentieth-century pathologists recognize the heightened danger of cardiac arrest in the minutes immediately after rising from bed.) Even after fixation, the stain of lividity may continue to darken, reaching its maximum intensity around twelve hours postmortem. It will remain prominent until overwhelmed by the colorful creep of bacterial putrefaction.

In the late 1700s, the French pediatrician and chemist Pierre Nysten revisited the postmortem muscle lock that the Greeks had dubbed rigor mortis. Like others before him, Nysten noted that death initially released all hold on the muscles. The body slumps, utterly flaccid, before gradually stiffening in the hours that follow death. He was the first, however, to build a clock out of the joint-by-joint progress of the subsequent rigidity. The result, in 1811, was the first scientific description of rigor mortis and "Nysten's law," which states: "The progress of cadaveric rigidity is descending." That is to say, it begins with the muscles of the face, then progresses to the neck, trunk, arms, and finally the lower limbs. Nysten concluded that the pattern reflected the increasing distance between different muscles and the brain, although he puzzled over the fact that decapitation didn't seem to affect the process. It is now known that the general progression of joint paralysis is actually from smaller to larger muscle groups. But that understanding—as well as an appreciation of rigor's many vagaries—would only come with the twentieth-century discovery of the biochemical wonders of muscle movement.

Meanwhile, the simplicity of Nysten's law gave early forensic pathologists a deceptively precise chart of time since death. The generally accepted timetable began with the first signs of jaw stiffness an hour after death and wrapped up with the lock of hips and knees ten hours later. The twelfth hour brought "full rigor," a fascinating state in which the body appears fully petrified. Rest the head on one chair and the feet on another, and the corpse will remain suspended like some bewitched volunteer in a vaudeville magician show. At typical room temperatures this rock-solid state lasts for twenty-four to thirty-six hours, before advanced decomposition begins to loosen the muscle groups in the same order that they seized.

About the same time that Nysten was refining his rigor chart, English physician John Davey became the first to thrust Gabriel Fahrenheit's 1710 invention—the sealed-glass, mercury-column thermometer—into a human body at autopsy. With this instrument, the students of death had been given the means to add hatch marks to the algor mortis clock, previously little more than a marker for the half day it took a corpse to become cold to the touch. Unfortunately, Davey began his experiments, not in his temperate homeland, but in the sweltering heat of Malta, the Mediterranean stronghold captured by the British in 1800. Consequently, the corpses of Davey's British soldiers actually rose as high as 108 to 113 degrees Fahrenheit in the hours after death. Nonetheless, Davey saw the implications—that the gradual equilibration of a body's temperature with that of its surroundings could be used to estimate time elapsed since death. Years later, he would become the first person to mention the forensic potential of cadaver temperature in his 1839 textbook *Researches, Physiological and Anatomical.*

Davey's widely read book prompted various pathologists to attempt their own algor mortis measurements. Unfortunately, the pathologists Davey so inspired failed to heed his wisdom in placing his thermometer *inside* the body. Temperature readings on the skin, typically in the armpit, resulted in what would become the near unshakeable dogma that body temperature dropped at a precise and steady rate of 1.6 degrees an hour (soon rounded to 1.5 for ease of calculation). Greatly popularizing this belief, in 1887 Frederick Womack published his to-the-minute, time-of-death calculations on 118 cadavers. Womack performed his temperature calculations in the mortuary and Anatomy Theatre of London's St. Bartholomew's Hospital, allegedly without being told the actual time of death recorded on each patient's death certificate. In the preamble to his first case, he apologizes for his "gross" error of estimating the patient's time of death at 4:54 P.M., when in fact the attending physician had recorded it as 5:05 P.M. Today, we know Womack's accuracy to have been impossible, given his or any other known method of estimating postmortem interval. Nevertheless, his reports solidified nineteenth-century beliefs in the pinpoint accuracy of algor mortis.

Together, the triple stopwatches of rigor, livor, and algor gave nineteenth-century pathologists the confidence to estimate time of death over the first twenty-four to forty-eight hours, up to the point where lividity became fixed, bodies reached room temperature, and rigor melted away. To expand their timetables into the long term, they turned to the scientific records of the previous century.

* * *

THE PIONEERING WORK of Europe's eighteenth-century anatomists led to the popular image of grave-robbing mad scientists, portrayed in Mary Shelly's *Frankenstein* and Charles Dickens's *A Tale of Two Cities.* It was no secret that many of the Continent's most eminent physicians participated in body snatching or employed their own "resurrection men" to supply their schools. Just how long these death-hardened scientists dared to study their rank cadavers before disposing of them remains unclear. We know they were familiar with the red, green, and black palette of bacterial growth that creeps across the body in the days and weeks after death. This *putrefaction*—the next important marker on the postmortem time line—brings an end to both rigor and lividity by turning muscle to mush and hiding lividity beneath a darker, broader stain.

The first outward sign appears as a subtle flush of green over the right lower abdomen, usually some twenty-four to forty-eight hours after death. Accompanying this emerald bloom is bloat, the postmortem distension that, in 1669, Sir Francis Bacon described as the work of "unquiet Spirits" fighting to break free of the mortal remains. Not until the 1860s would the "father of microbiology," Louis Pasteur, refute the "occult powers" and "internal agitation" of Bacon's unquiet spirits and identify the real culprits behind the bloat and stain of human decay. Though better known for his identification and eradication of food-spoiling germs, Pasteur sang the praises of the "moulds, mucors, and bacteria" that he claimed "combusted" the human body after death. Without them, he wrote, "life would become impossible because the restoration of all that which has ceased to live, back to the atmosphere and to the mineral kingdom, would be all of a sudden suspended." In other words, we'd all be knee-deep in carcasses.

Though Pasteur erred in assuming that the bacteria of decay required oxygen to "combust" the body (most, in fact, operate anaerobically), he set the course for later microbiologists to track the rate and direction of the bacterial growth in the hours, days, and weeks after death. Ground zero turned out to be the *cecum,* a cozy anatomical pouch at the head of the large intestine. During life, the body's normal bacterial fauna content themselves inside the cecum's warm puddle of liquid feces, with any overgrowth quickly beaten back by the patrolling white cells of the immune system. But once death garrotes their jailers, the cecum's bacteria multiply exponentially. Typically, their booming population breaks through the intestinal wall two to three days later. Spilling into the abdomen, they drift into the now passive circulation system, following its stagnant streams up the chest, down the limbs, and across the face.

The gases produced by the traveling bacteria mix and react with the deoxygenated blood to marble the body in shades of green and black along the feathery tracks of superficial vessels. At first, the delicate marbling remains confined to areas of livor mortis. By the fourth or fifth day, it tends to broaden into a black smear that shadows first the face, then torso and limbs.

Meanwhile, bacterial gases have inflated the trunk grotesquely. Tongue and eyes protrude. Lips swell and curl in an exaggerated pout. Breasts and genitals balloon to obscene proportions. Hair loosens and can now be pulled out in shanks. The internal pressure also forces bloodstained fluid from the mouth and nose—the benchmark purging described by Sung Tz'u in 1247.

It is at this stage that the once-smelled, never-forgotten odor of death permeates entire buildings and neighborhoods, making even the best-hidden corpse impossible to ignore. German chemists of the 1800s identified the pungent gases as the sulfur-rich waste products of protein-gobbling bacteria. Recognizing a new class of biochemical, they dubbed them *ptomaines*. Among the most prominent: the aptly named *putrescine* and *cadaverine*. The intensity of their repulsive odors—building, peaking, and subsiding in the weeks after death—provides its own strange clockwork.

Nineteenth-century anatomists were meanwhile documenting the "Swiss cheese" pattern of tissue destruction caused by the bubbling of bacterial gases through the liver and brain. Percolating to the skin's surface, these putrid gases likewise draw body fluids into dusky-colored blisters that loosen and lift the skin. Sometime between day three and the end of the first postmortem week, the loosened skin readily slips from fingers, hands, and limbs like fine-mesh stockings.

From such gruesome observation came forensic pathology's earliest textbooks, as well as the first institutes of legal medicine, established at universities in Leipzig, Paris, Lyons, Edinburgh, Glasgow, and London in the early to mid-1800s. At such centers, pioneers in the new field could carry out their research in close proximity to Europe's greatest detective forces, including those of Scotland Yard and the Paris Metropolitan Police.

Such collaboration between the worlds of medicine and forensics took longer to take hold in North America, where the politicized English coroner system, introduced in colonial times, still held sway. The situation would change with the sensational murder that shook Boston society just before the Christmas holidays of 1849. Fortuitously, both victim and prime suspect were intimately connected with Harvard Medical College, the former as a major benefactor, the latter a professor of chemistry. As such, both were friends with the country's leading anatomists of the day—medical men who volunteered their services in the murder trial that followed.

The story of Dr. George Parkman's murder has been retold many times in many ways, in literature, drama, and annals of the American legal system. In essence, the Boston socialite and philanthropist disappeared on the morning of his appointment with the quick-tempered professor of chemistry John Webster, a social climber known to live beyond his means. Also well known was the subject of the meeting: the long-simmering dispute over Webster having first borrowed money from Parkman and then duplicitously offering the agreed-upon collateral (a valuable gem collection) as security for a subsequent loan from another man.

When questioned after Parkman's disappearance, Webster presented a note apparently initialed by the doctor to acknowledge the loan's repayment that day. The two parted on good terms, Webster claimed. Support for Webster's version of events appeared in the form of witnesses who said they recalled seeing Parkman in town that afternoon. Police began to look for a common criminal who might have waylaid the doctor for the money.

Meanwhile, Webster's furtive doings inside his locked laboratory on the days following Parkman's disappearance raised the suspicions of the college's watchful janitor. In the middle of the night, the janitor broke through a basement wall to a pit beneath the laboratory's private toilet (little more than an indoor outhouse). There he found the decomposing remains of a partial torso, a thigh, and a shank of leg. Called to the scene, the city marshal ordered Webster's laboratory searched. In the furnace, the police found the charred remains of human bones, including a jaw.

In the sensational murder trial that followed, prosecutors faced the huge task of proving murder in the absence of a clearly identifiable body. Indeed, the case set legal precedent as to whether such a conviction was possible sans corpus delicti. By most popular accounts, the identification of the remains centered on the testimony of Parkman's dentist, who produced a denture mold that perfectly fit the charred jawbone handed to him by the prosecutor. The courtroom demonstration appeared even more compelling alongside a sketch of Parkman's profile with its distinctively protruding chin.

In reality, Webster's defense lawyers refuted the supposedly "perfect" match by calling their own dentist to the stand, a well-known denture maker who showed that the jawbone fit equally well in a half dozen dental molds from other patients. It remained unproven whether the jaw found in Webster's laboratory furnace was that of the murdered Parkman, or the leavings of some long-ago anatomy experiment. In fact, Webster's chemistry laboratory adjoined a dissection vault filled to the rafters with unidentified bones.

More damning was the testimony of the Harvard anatomists assigned to the prosecution's "postmortem committee." In his historic account of the trial, Simon Schama writes of Dr. Winslow Lewis's testimony on the state of the remains found in the laboratory privy:

> Though forensic, his language was quite different from the severe monotonies of the law; it explored and reported from the anatomy unsentimentally but with a kind of sensuous exactness, a poetic attention to hue, that riveted the attention: ". . . from left scapula to right lumbar region, of a dark mahogany color and hardened . . . a little greenness under the right axilla, probably from commencing decomposition, and some blueness under the left axilla—leaving the skin soft and easily broken . . . "

Dr. Lewis's unshakeable conclusion: When discovered by the college janitor, the remains in the privy pit had reached a state of decomposition consistent with death on the day of Parkman's disappearance. The jury found Webster guilty of murder, and he hanged.

The impact of the case and its grotesque medical testimony could hardly be underestimated. Warranted or not, the apparent precision of the expert's findings created an image of the doctor as infallible murder witness. Over the next twenty-five years, the United States moved rapidly to integrate medical experts into its antiquated coroner system, with one state after another amending its laws to authorize coroners to employ physicians to assist in their investigation of homicides and suicides. In 1887, Massachusetts took the next step, appointing the first state medical examiner, with supreme authority over "all dead bodies of such persons as are supposed to have come to their death by violence." The crime-ridden cities of Baltimore and New York followed suit, opening their own busy medical examiner offices in 1890 and 1915.

More important, interest in forensics was growing in academia, particularly at Harvard, where medical men continued to debate the sensational testimony of the Parkman murder. In 1937, the school established America's first program in "legal medicine" as a subspecialty of pathology, the postmortem study of disease. With the lavish support of the early Rockefeller Foundation, the residency-type training program became a greenhouse for the cross-fertilization of medicine and criminology, exposing the nation's most promising young pathologists to the expertise of its leading homicide investigators as well as world authorities in human anatomy and physiology.

Most promising of all was the opportunity for new research. At last, modern science seemed poised to transform the crude postmortem sandglasses discovered by the ancients into the precise medical chronometers needed to nail death to the hour, if not the minute. As the eager young bucks of pathology focused their microscopes and honed their histological assays on the tissues, fluids, and breakdown products of the dead, the message went out: Murderers beware.

With cruel irony, the next fifty years of research would instead reveal pathology's own version of Heisenberg's uncertainty principle. Though many scholars were loath to admit it, the closer researchers looked at the biological signposts of death, the muddier their markers appeared. Instead of sharpening the time-of-death determinations made by earlier pundits, their findings made any claim to accuracy look hopelessly naive.

2 Reasonable Doubt

Unfortunately, it is often the least experienced medical witness who tends to offer the most accurate estimates, not having seen enough cases to appreciate the many pitfalls and fallacies in the process.

—BERNARD KNIGHT, *THE ESTIMATION OF THE TIME SINCE DEATH IN THE EARLY POSTMORTEM PERIOD*

JUST BEFORE 11 A.M. on June 13, 1994, special investigator Claudine Ratcliffe of the L.A. County Coroner's office held an electronic thermometer in the air above another crumpled corpse. The digital dial at the end of the five-inch spike read 70 degrees Fahrenheit. Returning the device to her kit, Ratcliffe knelt beside the body, curled in a fetal position just inside the gate to a private walkway. She pressed her latex-gloved hand against the young woman's jaw. No movement. She tried shifting an arm, but it too remained locked in place, bent at the elbow and held close to the body. Lifting the short skirt of the victim's black dress, Ratcliffe noted a pale red stain across the underside of each thigh. She pressed her thumb against the discoloration and squinted to see if it would blanch. It did not. With a nod, she signaled the waiting officers to help her lift the body into the back of her van. In its privacy, she pulled the woman's skirt higher, exposing the belly and making a deft cut just inside the right rib cage. Into the short incision, she stabbed the same probe she had used to record air temperature five minutes earlier. From deep within the liver, the electronic sensor at the probe's tip registered 82 degrees.

Ratcliffe attached the thermoprobe's paper readout to her report for victim 94-05136, and noted "rigor mortis—fully established; lividity—fixed and consistent with position." The readings and notations exactly matched those already recorded for victim 94-05135, now lying inside an orange body bag on the other side of the van. Because it was her unfortunate duty to notify next of kin, Ratcliffe had likewise written down the names she'd found in her search of wallet and purse: Ronald Goldman, age twenty-five, and Nicole Brown Simpson, thirty-five.

All of ten minutes. That's how long it took Claudine Ratcliffe to record the triple markers of postmortem interval, or time since death, for two side-by-side homicides. But in the unprecedented marathon of a murder trial that followed, week after week of expert testimony would center on their meaning. The "Dream Team" of defense attorneys for the accused—football legend-turned-actor O.J. Simpson—derided the ineptitude of a coroner's office that could do no better than bracket time of death to between 9 P.M. June 12 and 12:45 A.M. June 13. Utterly useless. Witnesses could vouch for the whereabouts of both the accused and the victims as late as 9:30 P.M. Neighbors stumbled onto the bloodied bodies a few minutes after midnight, by which time Simpson was aboard a flight to Chicago.

This failure of the coroner's office to pinpoint time of death more precisely figured prominently in the Simpson defense team's argument that sloppiness, if not corruption, marred the entire case against their client. Did the killings occur within the one hour for which the defendant had no confirmed alibi, or did they not? Surely the county coroner's office could have determined that much, if only the police had summoned one of its investigators as soon as they arrived on the scene at 12:45 A.M.—instead of waiting nearly nine hours to do so. On cross-examination, the defense openly scoffed at Chief Medical Examiner Lakshmanan Sathyavagiswaran's insistence that he could not have further narrowed the time since death, had he himself beaten the police to the bodies.

Did Sathyavagiswaran actually expect the intelligent members of the jury to believe that medical science could do no better than his *barn door* of a time estimate? Understandably, the jury may not have followed the chief medical examiner's two-day-long commentary on the vagaries of time-of-death determinations. Or for that matter, his explanation of why the autopsies performed the day after the murders had not included what the defense was calling the new "paradigm" of postmortem clocks: a sampling of the eye's vitreous humor.

Incompetence? Or the limits of credible science? Truth be told, had Simpson's so-called Trial of the Century taken place in the nineteenth century instead of the twentieth, the medical examiner would have had a hard time justifying the broadness of his time estimate before a panel of his peers. But forensic pathology's time markers—so simple, sure, and seemingly exact in the 1800s—had begun to unravel by the turn of the century. By the 1980s, their scientific certainty had reached a nadir.

* * *

THE FIRST STANDARD to show signs of trouble was the pathologist's favorite stopwatch: algor mortis, or body cooling. The first truly scientific study of the phenomenon clearly identified its limitations. Unfortunately, its author, Dr. Harry Rainy, Second Regius Professor of Forensic Medicine at the renowned University of Glasgow, was not one to win kudos, neither among his peers nor in the public theater of the criminal court. In fact,

Rainy was an unlikely pioneer in the exciting new field of forensic medicine, just entering its heyday in the early to mid-1800s. An ophthalmologist by specialty, the aristocratic Rainy had been teaching "theory of physic," that is, medical science, at Glasgow for decades despite being passed over time and again to chair its medical department owing to his Tory leanings. When the Whig Parliament finally fell in 1841, the new Tory government quickly awarded the long-suffering professor the university's vacant regius, or "king's," chair in forensic medicine as consolation prize.

Rainy's lack of forensic expertise provoked stinging commentaries in both medical journals and local papers, which predicted—quite accurately—that the appointment would plunge Glasgow's budding department of forensic medicine back into obscurity. Indeed, several sensational murders took place in Glasgow during this time, including the trials of several socialites accused of poisoning their respective spouses and lovers. But the Crown looked elsewhere for its medical witnesses. Although the tall, handsome, and immensely dignified Rainy would have looked impressive on the stand, he was, to put it bluntly, "an insufferable ass." As further described by one of his pupils, Rainy was "not only cross-looking, but easily made cross. . . . Counsel quickly discovered they could make nothing of him in the witness box and so very soon gave him up completely."

But Rainy's snubbing left him ample time to pursue his exacting brand of dispassionate scientific research, something woefully absent from forensic medicine of the day. Most important, in the 1860s, he enlisted the help of Joseph Coats, a physician at nearby Glasgow Royal Infirmary, to carry out temperature observations on dying patients. By arrangement, Coats sent for Rainy whenever death appeared imminent in his ward. Together, they took temperature readings both on the belly and deep within the rectum, beginning in the moments before death and continuing at hourly intervals once they'd wheeled their subjects to the stony basement morgue.

At times, Rainy and Coats were forced to abandon their observations after as little as a half hour, their subjects too hastily collected for prearranged funeral services. But more often than not, they were free to continue their experiments through the night, before relinquishing the bodies to the unaware families sometime the next day. The occasional abandoned derelict, in turn, allowed them to record body temperature over several days, till the smell of putrefaction risked alienating the hospital staff.

With an eye for confounding factors that was far ahead of his time, Rainy wisely discarded all his body surface readings (too easily altered by a passing breeze) as well as fifty-four cases in which the mortuary temperature had been less than constant during the experiment period. Concentrating on the internal temperatures taken via the rectum, he then calculated the "ratio of cooling per hour" for each of forty-six cadavers.

In writing up his results for publication in 1868, Rainy became the first to evoke Newton's law of cooling to mathematically quantify algor mortis, at the same time pointing out the law's limitations when it came to describing the dead. According to Newton's law, a solid object loses heat at a rate directly proportional to the difference in temperature between the object and its environment. That is, a body that loses one-tenth of its heat between midnight and 1 A.M. would lose one-tenth of its remaining heat between 1 A.M. and 2 A.M., and so on. Accordingly, the algor rate would prove little use once the body neared room temperature—as it would enter a prolonged, near-level plateau before the temperature of the body and its surroundings finally became one. Before this point, it should be fairly straightforward to calculate time since death by extrapolating back from body temperature.

The problem was, Rainy's measurements showed that something else was going on during the first minutes to hours after death. While Newtonian physics would demand that heat loss be *fastest* during this initial period (when the temperature gradient between the newly dead body and its surroundings was greatest),

Rainy showed that initially, body temperature didn't budge, sticking at a "normal" 98 degrees Fahrenheit (37 degrees Celsius) for anywhere from several minutes to several hours. In one instance, he noted, body temperature actually rose in the first hour after death.

With startling insight for his time, Rainy went on to attribute this initial plateau and the occasional rise in early postmortem temperature to "physiological processes which are continued after the action of the heart ceases." That is to say, Rainy seemed to recognize that outward death—as discerned by lack of breath and heartbeat—was not the final word. Notwithstanding such complexities, Rainy was able to forge the first scientific formula for calculating time since death, albeit a complicated one. He wrote:

> Let the excess of temperature of the rectum at the first observation be t and at one hour later $t - t_1$: and let the excess of temperature of the body at death be $t_1 + D$, where D = the rectal temperature at death minus environmental temperature. Further, let $t/t_1 = R$ and, because t is greater than t_1 and consequently R will be greater than unity, then: log D - log t_1 divided by log R = X, which is the number of hours which have elapsed since death.

Rainy accompanied his formula with a cautionary warning that pathologists would be forced to relearn a century later. "Though we cannot calculate exactly the period which has elapsed since death," he wrote, "we can almost always determine a maximum and a minimum of time within which that period will be included." Specifically, Rainy stipulated a wide margin of error of about four hours and further warned against so much as attempting to estimate time of death by means of algor mortis in less than still air and steady temperature—in effect, eliminating most any corpse found outdoors.

Not surprisingly, Rainy's complicated mathematics, couched as they were in strict stipulations, drew little interest. Yet the same year, German pathologists independently confirmed his findings of an initial plateau in body temperature. They described it, however, not as some special attribute of the newly dead body, but as the physical consequence of insulation. In other words, any inert object with a low thermal conductivity—be it a heavy mug filled with coffee or a body layered in muscle and fat—will have such a "holding period" during its early cooling phase. Still, the Germans' numbers didn't quite add up. Insulation alone couldn't account for the duration of the plateau, even if one factored in the effect of clothing . . . nor could it begin to explain the sporadic cases in which body temperature in fact *increased* several degrees in the hours immediately after death. Regardless of the underlying explanation, the mere existence of an initial temperature "holding period" meant one thing: A corpse found at 98 degrees could be anywhere from four seconds to four hours dead.

Unfortunately, this hard-earned wisdom was utterly swept away with the sensational reports of London pathologist Frederick Womack in 1887. In his experiments at St. Bartholomew's Hospital, Womack used a special mercury thermometer with a flattened bulb of thin glass that he claimed could be read with an accuracy of "better than a fortieth of a degree." The claim seems ludicrous even by the standards of modern mercury-column thermometers. Moreover, Womack took his temperature readings on the cadaver's belly, attaching the thermometer to the skin with adhesive tape. The tape would have reduced the effect of cold drafts somewhat. Nonetheless, both the thermometer and the skin where it was placed would have been unduly affected by outside temperature. Not surprisingly, Womack saw no initial temperature plateau, because of his focus on the body's surface.

Instead, Womack reported a rock-steady temperature drop and, claiming pinpoint accuracy, recorded time of death to within minutes for over 100 corpses. Today, pathologists consider this utterly impossible. But at the end of the nineteenth century, it created a sensation. Amplifying Womack's renown was his position as a teaching fellow at the renowned St. Bart's in the heart of London. Indeed, he likely demonstrated his revolutionary new technique in the hospital's spectacular teaching wing, dedicated by the prince of Wales in 1879. Womack's autopsy room stood at the center of the new building, flanked on either side by two enormous teaching theaters large enough to accommodate hundreds of students, physicians, and visiting scholars.

Seemingly overnight, forensic doctors across Europe and the Americas embraced the idea that they could calculate postmortem interval with pinpoint accuracy. Dropping Womack's own rather complex mathematics, they settled on the deceptively simple formula still used by many pathologists today: that of adding one hour since death for every 1.5 degree drop below normal body temperature. Over the next century, this misleading bit of arithmetic would send countless murder investigations down cold trails, set free an unknown number of killers, and conceivably spell life imprisonment—even death—for a comparable number of innocents.

Compelling evidence of its naïveté resurfaced in the 1950s with a new flush of publications on cadaver heat loss. Carefully documented experiments by pathologists in Sri Lanka, Great Britain, and Germany resurrected the undeniable "lag period" that precedes postmortem cooling, as well as the occasional spike in temperature in the early hours after death. This time, science could explain the mysterious phenomena. But the explanation would fly in the face of the most persistent of all postmortem myths—that of the proverbial "moment" of death. Indeed, the very fact that pathologists still strived to pinpoint time of death—and judge and jury continued to accept their ability to do so—stemmed from this enduring concept of a singular instant when life snuffs out, like a house thrown into darkness by the slam of a master switch.

But a new definition of death was struggling to emerge in the 1950s—that of a progressive and drawn-out breakdown in the body's oxygen cycle. "Although the body cells require many different substances to carry on their life processes, the need for a continuous supply of oxygen is the most essential," former New York City medical examiner Milton Helpern first explained to lay readers in the *Encyclopedia Americana.* "Without oxygen, the brain, which normally uses about 25 percent of the blood's total oxygen supply, rapidly deteriorates. Other organs with high oxygen requirements soon also deteriorate and stop functioning."

But even after a living cell depletes its last dregs of oxygen, biochemists were discovering, it still refuses to go "gentle into that good night." Sensing the failure of its first-line power stations—the oxygen-dependent mitochondria—the cell falls back on anaerobic energy production, or fermentation. But the inefficient process of fermentation quickly depletes any stores of blood sugar, starch, and fat before forcing the cell to begin cannibalizing its own enzymes and membranes. Worse, without oxygen to complete their metabolic breakdown, the end products of fermentation convert to lactic acid. Slowly but surely, the acidity inside the cell builds to membrane-corroding levels.

All the while, this bodywide struggle for survival continues to generate warmth, though not as uniformly as the homeostatic humming of healthy tissue. Even the failing cell's last gasp—the moment when it either rips apart at the seams or, sensing its own crippled state, commits suicide—produces a tangible spark of biochemical heat.

Adding to this warming trend is the activity of any stray bacteria adrift in the ocean of salt water that bathes each cell, bacteria suddenly inundated by a feast of spilled cellular debris. In life, the immune system would send an army of white blood cells to mop up the mess and dispatch the gluttonous germs. But in death, the bacteria find themselves ignored, free to multiply exponentially and generate "body heat" of their own. Should a person die with an underlying infection—even something that would escape detection in life—the result can be a dramatic temperature spike.

As researchers began to understand this complex ebb and flow of life in the hours after death, they at last understood the vagaries of body temperature in the early postmortem period. "There are outposts where clusters of cells yet shine, besieged little lights in the advancing darkness," wrote the surgeon-poet Richard Selzer in 1974. "Doomed soldiers, they battle on. Until Death has secured the premises all to itself."

Their ignorance dispelled, forensic pathologists began reexamining the whole of algor mortis. Even if they steered clear of the uninformative plateaus at the beginning and end of algor mortis, they now realized they must face the complexities of how an irregular shaped, irregularly composed object lost heat. By recording temperatures at a number of different places both on and within the corpse, they confirmed suspicions that the rate of heat loss could vary widely, depending on where the reading was taken. Outside variables such as air temperature, wind, and humidity seemed to have the least effect on the brain, but the lack of easy access discouraged all but the most zealous practitioners. Most investigators settled on the next best thing, the rectal temperature, ideally taken deep within the bowels with a special, long-stemmed thermometer.

Even if they could agree on a standard location for cadaver temperature readings, forensic pathologists faced humankind's mind-boggling variability in size and shape, as well as proportional differences in fat and muscle. On top of this, they had to deal with the insulating effect of clothing—cotton versus wool, wet versus dry, layered, torn, and so on. Each variable affected the rate at which a corpse lost heat, as did the environment in which the body was found. A body lying atop pavement or tile, for example, would tend to lose heat more quickly than one on thick, insulating carpeting or grass. A breeze would accelerate cooling, whereas high humidity would have the opposite effect. Still more factors came into play when calculating cooling rates in bodies found in water, be it fresh, salty, or filled with sewage.

As the once simple formula of algor mortis devolved into chaos, the halls of pathology departments echoed with arguments over whether one must add blood loss to the equation. Though many scoffed, it seemed reasonable to wonder about what happens when a slashing victim loses quarts and quarts of 98-degree fluid. For that matter, did anyone really die at "normal" body temperature? Even ruling out fever stemming from sickness, something as commonplace as a victim struggling with his or her attacker could elevate temperature several degrees. In the 1960s, pathologists found additional evidence that a blow to the head could produce a dramatic spike in body temperature, stemming from damage to the heat-regulating center of the brain. Some scholars claimed that strangling victims experienced a similar sort of brain damage and temperature elevation from oxygen deprivation. Illicit drugs complicated the picture still further, with narcotics such as heroin lowering temperature before death and stimulants such as cocaine jacking it up.

Not that anyone was willing to throw in the towel on time-of-death determinations; they couldn't. The stakes were too high, and the pressure for answers from police and lawyers too intense. The best that forensic pathology could do was bolster the questionable accuracy of temperature-based death estimates with the lesser clocks of rigor and livor mortis. But these markers, too, lacked anything resembling a reliable time schedule.

The faint blush of early livor—first seen on the underside of a body between thirty minutes and three hours after death—had always been a subjective call, depending in no small part on the eyesight, even the imagination, of the death investigator. Moreover, anything darker than pale Caucasian skin made recognition increasingly difficult. Blood loss or anemia could likewise lessen lividity's appearance.

Still, hope of forging a time scale from lividity's changing color spectrum sprang from the discovery that blood's normal red color comes from the oxygen-rich pigment dubbed hemoglobin. As dying blood cells exhaust their oxygen reserves in the hours after death, their deoxygenated pigment darkens through shades of dusky red, purple, and blackish-blue.

Efforts to quantify the color shift proved difficult. Some of the world's leading forensic pathologists tried ordering their observations in lividity tables with such descriptive levels as "early," "well established," and "maximum intensity." But in the end, no two textbooks could agree on the beginning and ending times for each stage. Nor was there any way to standardize the perception of "faint" or "easily seen" for any two examiners. By the mid-twentieth century, most pathologists admitted that the variability of lividity's color shift defied precision.

At the least, researchers hoped they could forge a single marker for lividity's fixation—the point when its dusky patches no longer blanch to the touch. Debate over what exactly caused this fixation would continue into the twenty-first century. If it resulted from a hardening of the capillaries after body cooling, as many believed, then pathologists faced the catch-22 of confirming one temperature-based postmortem clock by using another.

Some claimed that the fixation resulted, not from body cooling, but from the permanent stain of deoxygenated hemoglobin as it penetrated the capillary walls under the pressure of gravity. Even accepting this temperature-independent cause, the fixation's predictability remained questionable. In 1964, German pathologists compiled timetables from leading forensic textbooks around the world. The resulting range for lividity's fixation—as early as four hours and as late as twenty-four hours after death—proved of little help in refining algor's one-day sandglass.

Meanwhile, the equally mysterious phenomenon of rigor mortis remained the subject of intense study. The key to its understanding had come with the discovery of adenosine triphosphate (ATP) in 1929. Soon after German biochemist Kurt Lohmann first isolated this energy-rich molecule in a ground-up matrix of muscle cells, it became clear that he had discovered the universal energy currency of life itself. The chemical energy stored in ATP's atomic bonds powered not only muscle contraction but also virtually every biologic function on Earth. ATP's postmortem role became apparent a half century later, and when it did, it became clear that the muscle lock of rigor—the pathologist's last best hope—could never be charted with accuracy.

The chemical changes behind postmortem rigidity resemble those that power muscle contractions in life. ATP fuels conscious muscle movement, but the lack of it triggers rigor mortis. During life, ATP energy breaks the chemical links that form between the muscle proteins actin and myosin. These are the same molecules that slide over each other during active muscle contraction. At the beginning and end of each muscle movement, actin and myosin join molecular "hands" with one another across adjacent muscle fibers. When muscles relax, the molecules release their mutual grip. As individual muscle fibrils exhaust their ATP reserves in the hours after death, they lose the ability to break the protein pair's handshake to allow free movement. So strong is the resulting chemical bridge between fibers that should one try to bend a large joint in full rigor, a bone may break before the muscle gives way. The same muscle freeze also affects involuntary muscles, often dilating pupils (sometimes one more than the other) and producing goose bumps, owing to the stiffening of the tiny muscle fibers surrounding each hair follicle.

At first blush, the straightforward chemistry behind rigor would seem the stuff of an ideal stopwatch. Even on a visible level, the postmortem stiffening appears to progress in an orderly fashion, moving from the smallest muscle groups to the largest. But the phenomenon's ultimate arbiter—the amount of ATP present in each muscle group at death—proved anything but predictable: The same fight-or-flight response that can elevate a murder victim's body temperature before death has an even more dramatic effect on muscle chemistry. First, the stress of an impending attack may actually boost ATP levels, as the nervous system gives the split-second signal to simultaneously raise blood sugar levels and shunt that energy-rich blood to the muscles. But any actual struggle or chase can quickly deplete the muscle's ATP gain.

Something as unknowable as the intensity of that fight, even the victim's aerobic conditioning, would then determine how completely the muscles drained their stores of fuel before death.

In life, the sprinter repays such an energy "debt" with gasping breath and racing heart. But in death, there is no pulse or pant. Muscle groups already depleted of ATP at death quickly give way to rigor. Indeed, the eerie phenomenon of "cadaveric spasm"—in which corpses are found forcefully grasping weapons, weeds, or even a lock of the assailant's hair—is now believed to result from such a near-instantaneous onset of rigor. By contrast, someone dying in her sleep, or for that matter struck unawares from behind, might take anywhere from one to seven hours to show the first hint of rigidity.

As a result of such vagaries, twentieth-century medical examiners found themselves desperate for better postmortem clocks at a time when the brave new world of molecular biology was revolutionizing the life sciences. In 1950, the world was abuzz with the fevered race to discover the structure of the mysterious molecule known as DNA, which many suspected to be the "blueprint" of life itself. American biochemist Linus Pauling had already revealed the helical structure of proteins, revealing their pivotal role in controlling virtually every biochemical process known. Catching the fervor, the world's leading forensic pathologists saw the tantalizing possibility of a body full of molecular switches and biochemical changes ticking off the moments since death with atomic precision.

Working hand in hand with leading biochemists, they tried charting the drop in levels of glucose and ATP and the simultaneous rise in the electrolytes and minerals spilled with each silent implosion of a dying cell. But the riot of cellular destruction that rippled through the body after death changed the chemistry of ordinary body fluids too rapidly to plot with accuracy. Then, in the early 1960s, several researchers independently discovered a most promising specimen source: the eye's vitreous humor. Eye jelly. A neat bubble of viscous, transparent goo sealed away from the rest of the body within the cavity of each eye socket.

In contrast to the rapid chemistry change of other body fluids, that of the vitreous humor unfolded slowly in the hours and days after death, as the dying cells of the retina slowly leaked their contents into the jelly-filled orb. Over the same time period, the vitreous humor gradually dehydrated, increasing the concentration of its postmortem dreck.

Researchers studying these changes considered a number of potential chemical markers, including increasing levels of nitrogen, sodium, chloride, calcium, and potassium within the postmortem eye. The latter proved the steadiest change, and in 1963, William Sturner of the Hungarian Institute of Forensic Medicine in Budapest published the results of his studies on vitreous potassium levels in fifty-four cadavers. Plotted against the first 100 hours after death, the values rose in a trajectory that Sturner boiled down into simple arithmetic—a formula no more complicated than that offered up for algor mortis a century earlier.[1]

A truly modern postmortem timepiece had been born, and it ticked for more than a day. Nearly a week. The news crackled through the world of forensic medicine, sparking overflow attendance for lectures and symposia at national and international conferences. The equipment needed could be found in any morgue—a needle and syringe to suction the eye. Most any hospital laboratory could then handle analysis of the specimen's electrolyte levels. Best of all, perhaps, the technique could be explained simply and clearly to a jury.

Suddenly, forensic pathologists were again stepping forward in the courtroom to ascertain time of death with authority. Lawyers rushed to keep up with the development, as juries tacitly ac-

---

[1] *(The formula, solved for hours [h] and using potassium concentration of millimoles per liter [K+] is as follows:* $7.14\,[K+] - 39.1 = 5.476 + 0.14\,h = [K+]$. *)*

cepted the findings as scientific gospel. Indeed, by the 1980s, defense lawyers were challenging the competence of medical examiners who *failed* to include a "vitreous potassium" in their autopsy reports—as did O. J. Simpson's attorneys in 1995.

Meanwhile, a few veteran pathologists were warning that, if they had learned anything from the past century, it was that in death, things are never so simple. Take a close look at Sturner's formula, they warned their colleagues. Sure enough, it described a straight line that bisected the scatter of potassium values from Sturner's cadaver eyes. But only half those figures actually fell close enough to this "statistical regression line" to represent an acceptable margin of error. Over the next two decades, a flurry of nearly two dozen studies tried to confirm and improve on Sturner's results. Most found even greater variation in postmortem vitreous potassium levels. Some even showed significant differences between the two eyes of the same corpse.

Nonetheless, interest in vitreous potassium continued to rise, along with the renown of its exactness. But even as the test reached its peak popularity in the mid-1980s, many of America's most experienced pathologists were packing away their syringes. Even more worrisome than the statistical variation recounted in the scientific literature was their own hands-on experience. The supposedly simple sampling technique of syringing the eye had turned out to be anything but easy.

Because potassium levels within the eye vary from front to back (where the retinal cells are located), the pathologist had to be certain he or she suctioned out every last drop of the vitreous humor to produce a reliable specimen. But pull just a little too forcefully on the syringe and you end up disrupting the delicate retinal tissue at the back of the eye. The result: a specimen contaminated with cellular electrolytes, most especially potassium—the very chemical they were trying to isolate and measure with precision. At best, such inadequate and spoiled samples could be discarded as useless. At worst, the pathologist could fail to recognize when a less-than-perfect technique produced flawed results and an erroneous time of death.

Finally, in the early 1990s, leading researchers concluded that vitreous potassium levels were only indirectly tied to time since death. The only direct relationship was with decomposition, specifically the decomposition of the retina. As a result, anything that accelerated postmortem decay—from increasing heat and humidity to a preexisting infection—could raise potassium readings to unpredictable levels. Unfortunately, many of those least qualified to recognize the limitations of the test continued to be tempted by its outward simplicity, a particular problem in the United States, where the vast majority of forensic autopsies outside large cities continue to be performed by nonspecialists.

Meanwhile, forensic science journals continued to report new experimental techniques. Most of the new tests centered around signs of so-called supravitality, or "life after death," in various tissues. In the 1970s, German pathologists had rediscovered a wealth of information on the subject from the late 1700s and early 1800s. The original research had resulted not from forensic interest, but public hysteria.

Since ancient times, people have worried about being mistaken for dead and then buried alive. Collapse and apparent death became especially common during the plagues that wracked medieval Europe. But at the dawn of the nineteenth century, sensation-mongering tabloids whipped such fears into an unprecedented fervor. Their reports of the "many ugly secrets locked up underground" included descriptions of claw marks seen on the inside of disinterred coffins. As a result, several renowned medical societies offered substantial rewards for scientific methods of ascertaining whether someone was *truly* dead. Among the prize-winning entries were numerous studies showing that muscles gradually lost their ability to respond to electrical stimulation in the hours after death and that eyes lost their response to a drop of "ergot" (a tincture of rye smut used to induce pupil contraction).

Repeating the experiments with modern methods, German, Swiss, and English pathologists confirmed their practicality in the 1980s, at the same time warning against subjective attempts to divide the strength of postmortem muscle responsiveness into a time scale (that is, "strong contraction" = one hour past death; "weak contraction" = six hours, and so forth). Rather, a simple "yes" or "no" checklist of postmortem reactions might help refine more classic time markers.

Putting it all together in 1988, Claus Henssge of the University of Koln published an unprecedented time-of-death worksheet. It included an artful "nomogram," or alignment chart, that Henssge had previously developed to partially rehabilitate the classic test of algor mortis. Henssge's truly amazing temperature chart incorporated parallel scales for a half dozen different variables such as body weight, air movement, and the thickness and layering of clothing. Arguably the world's leading expert on time-of-death determinations, Henssge had based the value of these "corrective factors" on experiments with thousands of cadavers of all sizes, in all manner of clothing, in still and flowing water as well as on dry land, at environmental temperatures from below freezing to 100 degrees Fahrenheit.

With his temperature nomogram alone, Henssge had found he could reliably bracket time of death to within six hours. By integrating the additional measure of muscle excitability and pupil reaction, he found he could sometimes narrow that window another hour or two without sacrificing reliability. Henssge's warning to would-be practitioners: "You can use all the right rules, but get the wrong results with less than precise information." Henssge even went so far as to list dozens of situations in which the pathologist should not even attempt time-of-death estimation, such as when the temperature around the corpse may have varied considerably over the postmortem interval or when there was a possibility that the victim lingered for some time before succumbing to the fatal injury. Indeed, the cautious Henssge was known to throw out as many of his own time-of-death determinations as he ever formally presented in German courts.

Throughout the 1990s, Germany's forensic pathologists reported great success in using Henssge's integrated methods. But others wondered whether German cadavers somehow retained the legendary orderliness of their culture. Try as they might, medical examiners in England, Switzerland, the Americas, and elsewhere were never able to achieve a comparable degree of accuracy with Henssge's technique.

So perhaps it was inevitable, in 1995, to find two of North America's leading medical examiners arguing over the quality and quantity of the mushy tubes of rigatoni found in Nicole Brown Simpson's stomach at autopsy. Did their state of digestion indicate that the victim had been killed within two hours of her finishing dinner, as the L.A. medical examiner suggested? or more than four hours after, as countered by Simpson's "Dream Team" expert, former New York City medical examiner Michael Baden. When challenged, both experts had to admit that the quantity and quality of stomach contents had long ago been dismissed as the most unreliable of all postmortem time scales. Such grasping at straws would continue to be part of medical expert testimony when all else failed.

Even as the O. J. marathon droned on inside Judge Lance Ito's courtroom, forensic science acquired its first textbook on postmortem timekeeping. Written by Henssge and his British counterpart, Bernard Knight, *The Estimation of Time Since Death in the Early Postmortem Period* was as much an urgent warning against the foolhardiness of certain knowledge as it was a state-of-the-art guidebook. "Little has changed from those early days [of centuries past]," wrote Knight in his introduction,

> Except that [our ancestors'] data acquisition equipment was merely the back of a hand to test the coolness of the corpse's skin, and their eyes and nose to evaluate decomposition. We now have multi-

> channel thermometry with thermocouples sensitive to a fraction of a degree, enzyme methods, vitreous chemistry, muscular reactivity and other avenues for collecting data. Regrettably, the accuracy of estimating the postmortem interval has by no means kept pace with the enormous strides made in technological sophistication.

Even with such modest ambitions, Knight and Henssge refused to venture beyond the "early postmortem period" delimited in their book's title. Past the first twenty-four to forty-eight hours after death, they warned, prudent pathologists must surrender any pretensions of science. That thankless wasteland—where bloat, rot, and maggots rob the body of outward identity—remained a forensic no-man's-land until well into the 1930s. Its eventual pioneers would come not from the world of medicine but a field better known for teasing out the eating habits of Neanderthals than the depravities of murder.

3 The Bone Detectives

. . . ah bone, is the pit of a man after
the cumbering flesh has been eaten away.

—RICHARD SELZER, *MORTAL LESSONS*

NOTHING CURED BILL BASS'S headaches so quickly as a call about a corpse. Even after twenty years of identifying human remains, the anticipation of a new case brought instant relief from whatever budget shortfall or stack of exams had started his head pounding in the first place. For that matter, Bass didn't think to grumble at being roused from his bed in the predawn darkness on a cold and rain-whipped morning just before New Year's—by all rights the midpoint in a college professor's winter break.

When Bass accepted the University of Tennessee's invitation to create an anthropology department in 1971, word quickly spread that one of the nation's leading "bone men" was about to take up residence. The governor's office immediately enlisted him to be Tennessee's first state forensic anthropologist, on call twenty-four hours a day. Having honed his art on the sparsely populated prairie around the University of Kansas, Bass expected to deal with the occasional discovery of skeletal remains. He requisitioned a van, filled it with archaeology tools, and assembled a forensic response team of eager young students.

But increasingly, Bass and his crew found themselves dealing with the recently dead. Frequently, the calls followed a fire from which the curled and blackened victims had been pulled, burned beyond recognition. At least that's how the police described them. "I don't think anyone can be burned beyond recognition," Bass told his students. It would be their job to gently simmer and scrub the charred bones clean, arrange them in anatomical order, and perform the precise measurements and detailed examination needed to determine stature, gender, age, ancestry, and telltale signs of prior injuries or anomalies that might be matched against medical records.

With half the land and twice the population of Kansas, Tennessee also coughed up its murder victims more quickly. A cornfield gully could hold its secrets for years, but no basement, car trunk, or plastic garbage bag could hide the smell of death from close neighbors for more than a few days. But by the time its fetid odor attracted attention, such a corpse had likely reached a stage of advanced decay that defied routine identification. For good reason, Bass's name could be found in the Rolodex of every homicide detective in the state. It didn't hurt that, with his flattop haircut, linebacker build, and easy laugh, Bass blended seamlessly with the usual crowd at any crime scene . . . save for the cluster of gangly, long-haired college students that usually accompanied him.

Bass's acceptance among law enforcement was surpassed only by his popularity with Tennessee's coroners and medical examiners, who could not believe their good fortune at finding someone willing to take their hated "stinkers." Not that they couldn't justify passing the buck, so to speak. Once bacterial putrefaction sets in, there is seldom much to be gained by standard autopsy.

As a result, Bass had been at the University of Tennessee less than a year when he appeared in the chancellor's office with what must have been the strangest request in the school's history. The new professor needed a place to put dead bodies—a very large place, preferably far from the air breathed by students and staff. Once recovered from his shock, Chancellor Jack Reese gave Bass an empty sow barn surrounded by acres of untended farmland near the head of the Tennessee River, where the school had once raised livestock.

Not long after, Bass traded in his van for a pickup truck with a passenger cab wholly sealed off from the camper shell where his archaeology tools now shared space with body recovery bags, latex gloves, surgical scrubs, and several gallons of bleach and disinfectant soap. Even a detached scientist of death needed some respite from the odor of his ripest subjects.

And it was a pungent corpse that Bass expected to be handling as he drove west out of Knoxville on a late December morning in 1977, without the help of his usual cadre of graduate students, all of whom had headed home for the winter holidays. Fortunately, so had Bass's son Charles, a graduate student in forensic anthropology at the University of Arizona, now sitting shotgun in the cab of his father's truck. There was not yet much to say about the work ahead. The previous afternoon, the Williamson County sheriff's office had responded to a complaint from the new owners of an antebellum mansion outside Franklin. During renovation, the couple had noticed disturbed soil in the Civil War–era graveyard behind the home, a tiny family cemetery belonging to the mansion's original owners. Suspecting vandals, the responding officers found something far more disturbing. It appeared that someone had tried to add a new occupant to the Shy family plot. Detective Captain Jeff Long immediately dialed Bass's home number, hoping for assistance in removing and identifying their newest "John Doe."

"Who is it?" That was the question Bass had been trained to answer, indeed the question around which his profession had been built over the past three decades. But the demands on Bass's expertise were changing. For starters, the rolls of the disappeared in a more populated region such as the Southeast were legion. So any hope of matching human remains to hundreds of missing person reports hinged on Bass narrowing the "when" of death and disappearance. Adding to the pressure to pinpoint time of death was the new camaraderie between police and forensic anthropologists such as Bass, who had left the isolation of their laboratories to become part of the investigation team at the scene of murder. There were times when all the scientific objectivity in the world couldn't disguise the cruelty Bass encountered, or buffer the burning questions from detectives who needed to retrace the steps of an unseen killer. "How long do you think he's been dead, doc?" Since coming to Tennessee, Bass had been asked this question countless times. And he'd gotten weary of hearing himself say, "I don't know."

But the dating methods of physical anthropology, the larger field in which forensics was the newest specialty, had been honed on the remains of ancient humans, going back over a million years in the fossil record. With a Geiger counter in hand, the anthropologist could assign fossilized bones to a particular millennium based on the known decay rates of radioactive minerals, or date more recent—that is, still organic—skeletons to a particular century with naturally radioactive carbon (C-14). Radiometric methods proved next to useless to date bones deposited since the 1850s, when the burning of fossil fuels began to skew the ratio of carbon isotopes in the atmosphere. And forget about carbon-dating the remains of anyone still breathing when the testing of nuclear weapons sent background radiation permanently off the charts in the 1940s and 1950s. Nonetheless, physical anthropologists had built an entire science on their ability to tease out the details of unseen human activities from dusty clues. If murder left some physical time marker in its wake, they had the ideal background to find it.

* * *

LIKE THE PRACTITIONERS of many uniquely twentieth-century sciences, forensic anthropologists can look back through antiquity to see the first glimmerings of what would become their profession. The oldest known example comes from the reign of the murderous Roman emperor Nero. In the year A.D. 62, Nero promised his ambitious mistress, Poppaea Sabina, the head of his newly divorced wife, delivered on a silver platter. Ever suspicious, Poppaea remained unconvinced that the head was indeed that of her rival until one of her courtiers pointed out the discolored tooth that had been the empress's least flattering feature. Though far from science, the gruesome identification epitomized the early work of forensic anthropologists: giving names to the dead through the study of variation in physical traits, most especially those of the bones and teeth, which persisted long after the flesh fell away.

As a scholarly pursuit, anthropology took root in early nineteenth-century Europe, a direct outgrowth of heated arguments over whether different human races constituted separate species or if all people could be traced to the biblical union of Adam and Eve. For evidence of the truth, each camp looked to variations and similarities in human skeletons, most especially the skull and purported differences in the volume of the cranium, or brain case. The direction of early research only degenerated with the advent of Darwinism, or evolutionary thinking. Across Europe, anthropologists began using skeletal measurements, or anthropometry, to support pet theories on the hierarchy of races, the superiority of men over women, even that of the upper classes over the poor.

Modern anthropologists look back on this era as the Dark Ages of their science. But from it would come the realization that skeletal variation could be used to reveal hidden identity. Questions of superiority aside, the bones of a man do tend to be thicker and longer than those of a woman, the pelvis deeper and narrower, the brain case larger overall but less protruding at the forehead. The bones of the face do vary across broad racial lines, particularly in their width, length, and forward projection. Evidence of age at death and occupation in life can be glimpsed in the number and condition of skeletal bones. A human comes into this world with some 450 separate pieces, which later fuse, either wholly or partially, to create the 206-bone inventory of an adult. The wear and tear of aging, in turn, leaves joints pitted and craggy, and trauma and habitual activities produce telltale scars and bony ridges.

By the mid-1800s, European anthropologists had compiled measurements from hundreds of skeletons, ranging from the tiny bones of stillborn babies to the brittle and bent remains of centenarians. Ironically, the first application of their knowledge to forensic science was not to identify faceless victims, but to brand criminal types. Italian anthropologist Cesare Lombroso, the controversial father of criminology as a science, believed in the "born criminal"—*l'uomo delinquente*—a kind of substandard human easily recognized by physical stigmata such as a bulging brow, large jaws, high cheekbones, and overly long arms. Criminals were evolutionary throwbacks in our midst, according to Lombroso, who argued that their distinctive physical traits came with an intrinsic "craving of evil for its own sake, the desire not only to extinguish life in the victim, but to mutilate the corpse, tear its flesh and drink its blood."

Bolstering his melodramatic theories with Darwinism in the 1870s, Lombroso won renown as a modern thinker. Criminal courts throughout Europe embraced his ideas. Before his reckless brand of criminal anthropology faded away in the early twentieth century, scores of men and women had been convicted, at least in part, on the allegedly "apelike" profile of nose and brow or the "prehensile" outlines of their feet. In some instances, Lombroso and the prejudices he inspired led directly to lengthy prison terms, even life sentences, for relatively minor offenses, justified on the basis of isolating inborn criminals from society before they could do worse harm.

Across the Atlantic, anthropology remained nascent, although considerable research on the human skeleton was being done by medically trained anatomists such as the Harvard professors who testified in Boston's sensational 1850 Parkman murder trial. Their famous identification of the slain Boston Brahmin from his distinctive jawbone can be found in the published histories of both forensic pathology and forensic anthropology. These same Harvard anatomists taught the country's first courses in anthropology, beginning in 1890. Four years later, George Amos Dorsey graduated with a degree in the field and went on to become curator of anthropology at Chicago's Field Museum of Natural History and, in 1897, the first anthropologist to take the stand in an American court of law.

The anatomical aspects of Dorsey's testimony had many parallels to those of his professors in the earlier Boston trial. But if the gruesome aspects of the Parkman murder set tongues wagging, those of the Luetgert case triggered a veritable gag response. The proprietor of a sausage factory in 1897, Luetgert was accused of murdering his wife and boiling down her diminutive corpse in an industrial steam vat—in and around which were found bony fragments that Dorsey identified as coming from a human temporal bone (skull), metacarpal (finger), and humerus (arm). Luetgert got life in prison. But Dorsey's testimony in the sensational trial, not to mention the fawning attention he received in the media, created a backlash in academic circles of the late nineteenth century. Criticized and ridiculed for stooping to courtroom theatrics and sullying "pure science" in the pages of tabloid papers, Dorsey soon left both anthropology and the country to become a U.S. naval attaché in Spain.

Although anthropology continued to flourish in North America, no respectable practitioner dared play a visible role in criminal investigations for another forty years. Meanwhile, the science split along two distinct paths. Cultural anthropologists, as popularized by Margaret Mead in the 1920s, focused on the traditions and folk beliefs of primitive societies. Physical anthropologists in the mold of Mary and Richard Leakey continued the work of tracing human evolution and variation through the study of bones across time and geography. It was this latter tradition that would eventually return to the service of forensics. But it would take a fortuitous real estate transaction of the 1930s to bring the two worlds together again.

In November 1932, FBI director J. Edgar Hoover announced the opening of the bureau's first crime lab in Washington, D.C., "Aroused by the ever-increasing menace of the scientific-minded criminal schooled in the fine arts and inventions of organized crime, the United States Bureau of Investigation has established a novel research laboratory where government criminologists will match wits with underworld cunning," reported the *Washington Evening News.* The concept of scientific crime-solving had already caught the public imagination, thanks to the successful use of ballistics, or bullet-marking patterns, in the prosecution of Prohibition era gangsters. In reality, the FBI's new Scientific Crime Laboratory consisted of little more than a converted lounge, chosen because it had modern plumbing and a sink. The fortuitous part was its location near the National Mall, directly across the street from the Gothic spires of the Smithsonian Institution.

It did not take long for Hoover's G-men to find their way across Independence Avenue to the Smithsonian's Division of Physical Anthropology and the laboratory of Ales Hrdlicka. Given the curator's thick Czech accent and eccentric European airs, the FBI agents had to surmount more than the usual cultural divide between law enforcement and academe. But at Hrdlicka's fingertips was the Smithsonian's vast reference collection of human skeletons, well on its way to becoming the world's largest. His own position funded by federal dollars, Hrdlicka felt compelled to acquiesce to what would become a Smithsonian tradition of quiet cooperation, supplying clues to the identity of human remains—that is, decomposed murder victims—but avoiding either court testimony or written record of his forensic work.

It would take a physical anthropologist outside the Smithsonian's hallowed halls to break the final taboo. A no-nonsense giant of a man teaching anthropology at the University of Pennsylvania, Wilton Krogman inherited a fair amount of the overflow FBI casework from his Smithsonian colleagues. Seeing the greater potential for public service, in 1939 he published the first "Guide to the Identification of Human Skeletal Material," describing in explicit terms what physical anthropologists could do in the service of homicide investigation. Not that any scholarly journal would touch the paper. Krogman's landmark article came out in the FBI's *Law Enforcement Bulletin*. The affront could be forgiven, perhaps, as it drew little notice outside law enforcement circles. Still, university anthropologists across the country were known to curse Krogman for the new intrusion of police detectives knocking at their doors.

As World War II drew to a close in 1945, a sense of patriotism drew scores of physical anthropologists to the work of identifying the skeletal remains scattered across the battlefields of Europe and the South Pacific, work they would continue in the 1950s with the Korean conflict. Finally, in 1962 the budding field of forensic anthropology got its first bona fide celebrity, a Smithsonian curator who took an almost childlike delight in the crime-solving aspects of his work. Despite the condescension of colleagues, Lawrence Angel welcomed the attention of reporters nosing around for a good story. Balding, with muttonchop whiskers and colorful bow ties, the irrepressible Angel did not at all mind the moniker "Sherlock Bones" appearing above his picture in magazines ranging from *Science Digest* to *People*. The national publicity sparked a dramatic increase in awareness among even small-time police departments and coroner's offices. Beginning in the late 1960s, Angel established a tradition of fall workshops on physical anthropology for coroners and medical examiners, in essence educating them on the services his field could provide them. By his own estimate, Angel's workshops eventually reached over half the medical examiners in the country.

Information and interest began to flow the other way as well. In the late 1960s, a handful of anthropologists began attending annual meetings and workshops held by the American Academy of Forensic Sciences (AAFS). In 1972, the academy established a special section devoted to anthropologists. In this heady atmosphere, the term "forensic anthropologist" was first set down on paper and the first generation of anthropologists to openly dedicate themselves to the art honed their skills. Among them was Bill Bass, who earned his doctorate in physical anthropology under the auspices of the legendary Krogman in 1961.

Still, Bass found that his world-class apprenticeship and long personal conversations with Krogman had not begun to prepare him for the time-of-death questions that would come to dominate his forensic work in Tennessee. Nor was he alone. Fellow pioneers such as Larry Angel at the Smithsonian, Clyde Snow at the University of Oklahoma, William Maples at the University of Florida, Sheilagh Brooks at the University of Nevada, and Jane Buikstra at Northwestern University faced the same pressures to answer the "when" as well as "who" of murders. Trained as they were to view flesh as myth, an ephemeral if distasteful obstacle to the solid reality of bone, they came ill prepared to offer any schedule for its gradual disappearance over the months to years after death. Nor were they inclined to do so.

Unlike the medical pathologists who proceeded them into the world of forensics, the anthropologists largely resisted the temptation to offer more than they knew with scientific certainty. Perhaps the difference lay in their training. Medical schools may prepare their students to take their place as society's unquestioned demigods, but the peer-review system of scholarly science is quick to crush students who offer results unsupported by verifiable fact and controlled methods. So although vast collections of skeletons gave anthropologists the confidence to describe stature, gender, and age at death, they lacked any such reference set for the journey from death to dust. A detective arriving at their doorstep with a bag of dry bones could expect nothing more exact than, say, "six months to sixty years."

Still, the busiest practitioners of this new science, people like Bass who handled upward of a hundred forensic cases a year, couldn't help but become imprinted by the color, smell, and consistency of the bodies and bones they handled. When time of death later became known (owing to confession or other evidence), the impression stuck. Admittedly, certain time markers were harder to ignore than others. Angel's staff at the Smithsonian never quite got used to his habit of leaving his laboratory window slightly ajar when processing his most fragrant cases. Jokingly, the curator often quipped that the elapsed time from his cracking the window to the first complaint would yield a fairly accurate formula for how long someone had been dead.

More subtle observations likewise suggested, if not a strict timetable, then a relative sequence of change: Skeletons disarticulate in a more or less sequential pattern, ligaments giving way in smaller joints before larger ones. Bare bones gradually lose their greasiness as their fatty inner marrow degrades and disappears. Subsequent years and decades of exposure bring progressive bleaching, flaking, and cracking that gradually reduces bones to dust.

But the speed with which a body passed through such transformations, anthropologists quickly saw, could vary widely with both geography and circumstance. In a warm, humid region such as the Southeast, for example, an unburied body could take anywhere from two weeks to eight months to melt down to cartilage and bone connected by sinewy ligaments. In the same environment, complete skeletonization, marked by the total disappearance of cartilage and ligaments, suggested that a body had been lying exposed for longer than eight months. Bodies in many northern states and provinces followed a similar timetable in late spring and summer, with decomposition slowing considerably in the cooler months of fall and winter.

Burial, in turn, changed everything. Even a foot of dirt cover could double the time for a body to skeletonize. However, the same soil could speed the subsequent disintegration of bone. The more acidic the soil, the more rapid the breakdown, as low pH speeds chemical reactions that transform the bone mineral hydroxyapatite (a calcium phosphate) to crumbly calcium salts and phosphorous. As a result, any anthropologist venturing into the tricky realm of dating remains had to develop his or her own regional, and highly personal, set of signposts for the journey from death to dust.

Yet experience also taught forensic anthropologists to expect the unexpected. They swapped stories such as that of Maine anthropologist Marcella Sorg's buried double homicide: a shallow grave in which the uppermost victim was pulled from the ground completely skeletonized, whereas the corpse directly beneath emerged with its flesh wholly intact. The two men had similar builds, similar clothing, and similar bullet holes to the head. Detective work would later reveal that they had been shot within minutes of one another. Then what would explain the difference in decomposition? Perhaps the overlying body somehow protected its companion. Perhaps the seepage of its body fluids contributed to the remarkable preservation of the second body.

Even more common was the discovery of half-skeletonized bodies—wholly intact below some line where water, soil, clothing, or shade protected the flesh from degradation, with nothing but bleached bones above. How could anyone quantify the variables that produced such a picture?

What became clear was that such variability only increased with time since death . . . that and the impossibility of isolating the effects of any one factor without considering all the others. One subtle twist of circumstance—a change in temperature, humidity, sun, or soil pH—and, like a kaleidoscope, the tumbling pieces produced an entirely new picture.

Painful experience also taught the forensic anthropologist to beware of the subtle prejudices that came with knowing too much. Detectives confiding their suspicions had a way of filtering the scientist's vision until he or she saw just the right amount of flesh clinging to bone to place suspect and victim together at the right time. Then, a new piece of evidence, an unexpected confession, and the brazen judgment came back to bite. So it was that Bass did not stay on the phone too long when Detective Long called to ask his help that stormy December day.

Four hours after they left home in the predawn darkness, Bass and his son pulled into the rain-slicked parking lot of the Williamson County sheriff's office. Long jumped into the cab of the pickup for the drive up the Harpeth River along historic Del Rio Pike Road. As they wound their way past horse farms and a scattering of million-dollar homes, Long explained that the disturbed grave in which they found the body was that of Lieutenant Colonel William Shy, identified by a small granite marker as killed in the Battle of Nashville, 1864. Since talking to Bass on the phone, Long had assigned a crew of deputies to enlarge the muddy hole over the grave to gain access to both the body and Shy's cast-iron casket, to see if it had been damaged.

Arriving at the mansion, the three men pulled on heavy jackets and rain slickers and hiked down to the tiny family plot, on a slight rise near an old riverbank behind the Shy mansion. Peering through the misting rain, Bass saw his first clue as to how long the victim had been dead. The cluster of deputies nearest the grave stood well back and upwind. Despite near-freezing temperatures and a sheet of plywood placed over the grave, the reek of decaying flesh was overwhelming.

What a clever way to get rid of a body, Bass mused. Dig over a grave, dump the stiff, and get out before anyone notices you. Chances are, no one will ever find it. If they do, who's to think it doesn't belong? That is, if it goes undiscovered for long enough to blend in with the graveyard's other "occupants."

There was nothing to suggest the remains of a Confederate hero in what Bass saw when he pulled away the grave's temporary cover. At the bottom of the muddy pit, about three feet deep, was the hulking form of a man dressed in a tuxedo, or at least what was left of him. The corpse had no head, a grisly but not uncommon find in Bass's forensic work.

As most of the bystanders took a giant step back, Bass slid into the muddy hole, just large enough for him to plant his feet on either side of the corpse. Clearing away more mud, he could now clearly see the shreds of a tuxedo shirt, a vest, an expensive-looking coat, and emerging from a shirt sleeve, a white-gloved hand. The clothing immediately evoked the society pages of the local paper, this being Tennessee walking horse country, or perhaps a maître d' in a pretentious restaurant, thought Bass. Most striking of all, for Bass, was the pink hue of the corpse's flesh.

The anthropologist thought back several years to a summer spent assisting the Tennessee Highway Department in relocating the Civil War–era cemetery of a German Lutheran community near Wartburg. Taken together, the dry and disintegrated remains of all nineteen graves could have fit in the palms of his hands. He doubted whether anything even remotely bodylike could be left of Colonel Shy.

Shifting back to the present, Bass noted that whoever dug the hole had apparently stopped when they hit the cast-iron coffin, gashing a hole into its rusted lid in the process. The body appeared to have been dumped feetfirst, in a squatting position atop the casket. As it began to decay, parts of the lower torso and legs had dropped in through the break in the coffin lid. Over the next four hours, Bass twisted and strained in the tight confines of the muddy pit to extract the remains. The limbs had separated from the pelvis and shoulders. But much of the torso remained intact, as well as the legs and the arms below the elbows. Bass handed the pieces up to Charlie, one section at a time, then climbed out to help him arrange them in anatomical order on the plywood plank next to the pit. The body was clearly male. Bass estimated that the man had been twenty-five to twenty-eight years old and weighed approximately 175 pounds at the time of death.

With a pocketknife, Bass cut into the pink flesh of the thigh to assess its firmness. Opening the shirt, he found recognizable portions of both the small and large intestines. Based on the amount and consistency of the decaying flesh and its overwhelming odor, he gave Long an estimated time since death of six to twelve months.

Finally, Bass wanted to look inside the casket. There was always the chance it contained the missing head. But the muddy excavation was too narrow and the hole in the casket too small for him to bend over on his hands and knees. With a chuckle, he realized there was only one position that would work.

As the disbelieving deputies struggled for footing in the mud, Bass had them suspend him headfirst over the casket, motioning them to lower him until his head disappeared into the gash in its lid. Shining a flashlight beam first to the head of the box and then the foot, Bass saw what he had expected. Just out of reach at the top of the casket, a couple of inches of disintegrated debris were all that was left of Colonel Shy. "This can wait," Bass called up. Breaking the grimness of the foul day and fouler odor, the younger men roared with laughter as they hauled the professor out of the mud and onto the grassy knoll.

It was late, the Friday evening before New Year's, and everyone wanted to get home. Bass offered to drive to the state crime lab in Donelson to drop off the victim's clothing along with two cigarette butts he'd found in the grave, presumably tossed over the edge by whoever dumped the body. For good reason, the laboratory staff wouldn't let him enter the building with his load. As the sun set and temperatures plummeted below freezing, Bass, his son, and a miserable crime tech sorted the rank material in a loading-dock driveway, repacking it in plastic evidence bags.

Wet, exhausted, and mud-spattered from head to toe, father and son headed home, showering and collapsing in bed just after midnight. Outside, in the locked shell of the pickup truck, the neatly bagged and divided remains of "John Doe" would keep until after the holiday weekend. The news, however, did not. The *Nashville Banner* recounted their exploits on the front page of the morning paper:

> Dec. 31st, 1977, Franklin. Williamson County Authorities investigating the tampering of a Civil War soldier's grave discovered that a second body had been placed over the grave probably within the last year. The body is an adult male, clad in what appeared to be a tuxedo. The body of Colonel Shy, in its steel vault, was undisturbed, officials said.

Bass didn't need an alarm clock to wake him early on Monday, January 2. With Charlie in tow, he pulled into the deserted parking lot of the University of Tennessee's Neyland Stadium just as the sun was lighting its rim. A massive coliseum of a football arena and the famed "Home of the Vols," the college bowl held darker secrets beneath the horseshoe bend of its south bleachers. When Bass arrived at the school to create an anthropology department in 1971, administrators had offered him his choice of some old houses or what had been a four-story, 166-room dorm for athletes built beneath the stadium seats. He chose the curving stadium's catacombs. Bass had always found it suitably mysterious that one could never see the end of a hallway, or for that matter who was ten feet ahead around the hallway's never-ending bend.

In the stillness of his laboratory, Bass and his son worked quickly and quietly. There wasn't much the professor could do with soft tissue all over the bones. Yet he didn't dare cut too close for fear of marring the nicks, knobs, and other minutiae that could reveal identity as well as manner of death. So they filled a restaurant-supply kettle and a turkey-size roasting pan with water and set them across the burners of the laboratory's gas range. To each Bass added a dash of Adolph's Meat Tenderizer and a half-cup of Bizz Detergent with Bleach. By 8 A.M., Bass was lowering the burners to get the gentlest of simmers—just enough heat to steam away the flesh without boiling off the cartilage that capped each joint. It would take hours. He didn't mind. But by noon, he began to sense something was wrong. The soup didn't smell right.

Not that Bass could place the odor. It wasn't the detergent or meat tenderizer; he knew those aromas as if they were his own. It was like something out of a chemistry experiment. Bass, who had never "processed" an embalmed body before, had a sinking feeling that this might be what one smelled like.

A call to the state crime lab added to his suspicions. Analysis had failed to find any trace of synthetic fiber. The coat, pants, and vest were cotton; the tuxedo shirt, silk. Meanwhile, Bass found a reference to nineteenth-century embalming methods. Arsenic. They used arsenic as an all-around pesticide and preservative for everything from cowhide to corpses. Late in the afternoon, Detective Long called to report on what they'd found when they pulled Shy's casket out of the ground: a jawbone, bone fragments, and what looked like a full set of teeth. He'd have them delivered in the morning.

Meanwhile, father and son had laid out the rest of the boiled down skeleton. The short, narrowly angled pelvis confirmed the professor's initial impression that the body was male. The diameter of the rounded head of the femur, or thighbone, likewise fell within the normal range for an adult male, as determined by Krogman in the 1960s. Femur lengths of 492 millimeters (right) and 492 millimeters (left) confirmed Bass's field analysis of someone just under six feet. Close examination of the joints showed that their bony caps, or epiphyses, had completely fused with the bones' main shafts, but without obvious signs of wear and tear—likewise consistent with an age of twenty-five to twenty-nine.

That afternoon the sheriff's department released Bass's description to the media—"white male with brown hair, approximately 5'11", 175 pounds, dead six to twelve months"—along with the county sheriff's plea for information: "It looks like we have a homicide on our hands." But Bass already had his doubts. He looked at his scribbled transcription of the words on Shy's tombstone:

LT. COL W. M. SHY
20TH TENN.
INFANTRY C.S.A.
BORN MAY 24, 1838
KILLED AT BATTLE OF
NASHVILLE
DEC. 16, 1864

Twenty-six years old. Bass's inquiries with local historians turned up further details about how the young lieutenant colonel had met his end: A musket ball to the brain, at close range, "his head being powder-burned around the hole made by the shot."

When the skull fragments arrived at the lab, the final pieces of the puzzle fell into place, all seventeen of them. Brown, almost chocolate in color, the fragments fit together to form a complete skull, intact except for a near-perfectly-round twenty-millimeter hole about two inches above the left eye and a larger, more jagged exit hole at the back of the skull. The force of the bullet had been so great that it shattered the skull into seventeen pieces. Bass recognized their color as that of bones that had been buried for decades. Examining the teeth, Bass found an abundance of deep cavities but not a single filling. Fitting them into place left several gaps, suggesting extractions. That alone might have alerted him, Bass thought with twenty-twenty hindsight. Dentistry did not evolve much beyond tooth pulling till the late 1800s.

The skull, like the skeleton laid out before him, no doubt belonged to Lieutenant William Shy. Soon after informing the Williamson County sheriff of his stupendous mistake, Bass fielded the expected phone call from the *Nashville Banner.* "I got the age, sex, race, height and weight right," he told the reporter gamely. "But I was off on the time of death by 113 years."

Over the following week, as his graduate students drifted back to school, Bass enlisted their help in ferreting out more detail behind the most baffling aspects of the case. Historical accounts confirmed that Shy's family had spared no expense in having his body thoroughly embalmed and dressed in civilian finery before entombing him in a top-of-the-line lead coffin. Perhaps the sturdiness of that coffin had something to do with the remarkable preservation within. By contrast, Bass's previous experience with Civil War–era remains (the historic Lutheran cemetery) had involved unembalmed bodies in thin pine boxes. As for the advanced decay of Shy's skull, his musket-shattered head would have been the one part of his corpse unable to hold embalming fluids. Indeed, its putrid decay would have begun on the battlefield.

When local police resumed their investigation of the disturbed grave, neighbors near the mansion admitted they weren't surprised to hear of the vandalism. Though no one would name names, there had been a spate of recent interest in Confederate memorabilia. Specifically, the word had been circulating that (a) Confederate colonels were *always* buried with their swords and (b) such an artifact would be worth tens of thousands of dollars.

Yet for all the trouble that the looting had caused, it seemed doubtful that the looters had come away with anything more than a few nineteenth-century buttons. Local historians concluded that Shy's sword had most likely been stolen over a century earlier, as he lay dying on the battle hill that came to bear his name.

Still, the case continued to haunt Bass. More than a humorous anecdote or a cautionary tale against attempting a rushed analysis in the field on a miserable day with muddy specimens, it seemed to epitomize a glaring gap in his science. Already, the professor could hear the words of the next attorney to cross-examine him on the witness stand. "Isn't it true, Dr. Bass, that you have been known to be over a hundred years off in your determinations?" In fact, given Bass's preeminence in the field, his mistake could be used against *any* forensic anthropologist offering however broad an estimate of time since death. And the challenge would be entirely justified, Bass concluded. If anything good was to come from his blunder, Bass thought, it would be for it to spur desperately needed research.

Over the following months, time since death came to dominate more and more of Bass's class discussions and after-hours gab sessions with grad students. He scoured the scientific literature to find anything on the subject. Textbooks on forensic pathology summarized early postmortem changes such as body cooling, rigor mortis, and lividity, but generally left off where Bass and his recovery team had to pick up—at the point where putrefaction robbed the body of outward identity.

In 1962 Bass's mentor, Wilton Krogman, had given time since death barely a nod in forensic anthropology's first textbook, *The Human Skeleton in Forensic Medicine.* Four years later, an anthropology student at the State University of New York described the relative decay of 554 skeletons from a Seneca Indian cemetery in her doctoral dissertation. It was the kind of work that grabbed Bass's interest. But the Seneca graves were decades to centuries old. Forensic anthropologists needed research on more recent remains.

Bass saw little progress in 1979, when Smithsonian curator Dale Stewart published his long-awaited *Essentials of Forensic Anthropology,* the first textbook to update Krogman's pioneering work. Stewart devoted only 7.5 pages of his 300-page primer to the cycle of human decay, concluding "there is no escaping the fact that, for most skeletonized remains, estimation of time since death usually is little more than an educated guess."

Despite the dearth of research, the topic continued to generate tremendous interest in forensic circles. Any presentation even remotely touching on the subject drew overflow crowds at meetings of the American Academy of Forensic Science. The 1980 AAFS conference featured a pair of such papers. One described two years of research by Florida State anthropologists and the state's Department of Law Enforcement on dating human remains by the deterioration of associated clothing. As part of their field training program for homicide investigators, anthropology professor Dan Morse (a retired pathologist) and archaeologist Jim Stoutamire had their students scatter and bury dozens of fabric swatches in a half-dozen different kinds of soil from clay to sand. They then systematically began to record the materials' decay at monthly intervals over four years.

Buried rayon and cotton proved the quickest to disintegrate, with damage apparent to the naked eye within weeks, total deterioration in fifteen to seventeen months. Silk and wool, both buried and on the surface, showed signs of microscopic damage at ten months, with damage becoming visible to the naked eye at fifteen months, followed by total destruction at thirty-five months. Leather and synthetic fabrics proved the most resistant over time, showing no visible damage after two years. Sunlight degraded some fabrics such as acetate more quickly than it degraded others such as acrylic. Soil type and moisture likewise influenced the rate of destruction. Cotton and rayon, for example, broke down within ten months when buried in wet, swampy soil, but took twenty-five months to degrade in dry clay and sand. By contrast, moisture actually increased the durability of wool and silk. Bass paid particular attention to the study's finding that temperature proved to be the most important environmental factor by far. Heat accelerated both chemical degradation and microbial destruction. The microorganisms capable of destroying many fabrics appeared to operate best at temperatures above 70 degrees Fahrenheit—confirming what Bass had seen in many of his own cases.

At the same conference, Bass listened to a University of Illinois anthropologist describe how plants and fungi speed decomposition, splitting joints and penetrating bone with their tendrils and roots. Both studies reflected what Bass considered the strength of forensic anthropology—approaching the death scene like an archaeologist who never underestimates the importance of the smallest artifact, some tiny clue that might prove the key to unlocking unseen lives . . . or deaths.

Bass's own early research on the burial practices of South Dakota's Arikara people reflected the same crucial focus on what might be called the *ecology* of death. In the summers of 1965 and 1966, he had led a field party excavating hundreds of graves from the site of a 165-year-old village once visited by Lewis and Clark and later shelled by the U.S. Cavalry. As Bass and his students began shoveling and chiseling the prairie clay from the village graves, they noticed what at first glance appeared to be seeds filling the eye sockets and pelvic cavities of many skeletons. Still more of the mahogany-brown pellets poured out of the buffalo robes that swaddled the bones. It was then that the anthropologists realized they were looking at some sort of insect cocoon.

On returning to campus, Bass showed the tiny artifacts to his colleagues in the entomology department. They identified them as the pupal cases of flies, most likely an array of the blow flies and flesh flies that sweep north into the Dakotas as soon as the temperatures rise above freezing each spring.

Bass immediately saw the implications. He and his students could use the puparia to identify the season of death. Graves containing puparia would have been dug and filled between April, when the flies arrived each year, and mid-October, when the first hard frost wiped out the year's brood. Graves free of puparia would correlate to deaths in late fall and winter. Bass and one of his graduate students, Miles Gilbert, went on to publish a paper on the possibility of using fly puparia to not only seasonally date burials, but potentially bracket periods of epidemic, war, or starvation.

With those pupal cases, Bass had unknowingly stumbled upon a science that could supply the postmortem timepieces he so desperately need in his later forensic work. Admittedly, he would not have found a single citation describing the forensic use of insects in a modern scientific journal. For that matter, this science had yet to be given so much as a name in North America, despite its deep roots in the Old World.

4 THE WITNESS WAS A MAGGOT

We fat all creatures else to fat us, and we fat ourselves for maggots. Your fat king and your lean beggar is but variable service—two dishes, but to one table. That's the end.

HAMLET (ACT 4, SCENE 3)

NOT UNTIL THE 1980s would an American entomologist add the line "Forensic Consultant" to his curriculum vitae. Yet, whenever modern-day forensic entomologists step before an audience—be it a jury, college class, or a room full of homicide detectives—they invariably introduce their science as "ancient," nearly 800 years old. They trace its first known use to a tale of murder by slashing recorded in Sung Tz'u's thirteenth-century Chinese detective manual, *Hsi Yuan Chi Lu* (The Washing Away of Wrongs).

On a sweltering afternoon, a group of farmers returning from their fields outside a small Chinese village found the slashed and bloodied body of a neighbor by the roadside. Fearing bandits, they sent for the provincial death investigator, who arrived to convene an official inquest. "Robbers merely want men to die so that they can take their valuables," he informed the gathered crowd. "Now the personal effects are there, while the body bears many wounds. If this is not a case of being killed by a hateful enemy, then what is?" Nonetheless, questioning the victim's wife revealed no known enemies, at worst some hard feelings with a neighbor to whom her husband owed money. On hearing this, the official ordered everyone in the neighborhood to bring their farm sickles for examination, warning that any hidden sickle would be considered a confession to murder. Within an hour, the detective had seventy to eighty blades laid before him on the town square. "The weather was hot," Sung Tz'u notes. "And the flies flew about and gathered on one sickle," presumably attracted by invisible traces of flesh and blood.

Its owner turned out to be the neighbor who had lent the dead man money. At first, he denied anything to do with the murder. Then the investigator forced him to look at the flies, which now covered his farm tool, despite its outwardly spotless appearance. "The sickles of the others in the crowd have no flies," the official pronounced. "Now you have killed a man . . . so the flies gather." According to Sung Tz'u, "the bystanders were speechless, sighing with admiration. The murderer knocked his head on the ground and confessed."

Arguably, our unnamed thirteenth-century detective demonstrated more sagacity than science, and time since death played no role in his entomological sleuthing. Nevertheless, the tale perfectly illustrates the "ground-zero" from which forensic entomology would build an unprecedented postmortem clock. Certain flies find the smell of death irresistible. Within minutes, sometimes seconds, they materialize as if from thin air. Their goal: to deposit their eggs on the still-tender flesh that alone can brood their young. Anyone who can gauge the speed at which these larvae, or "maggots," develop can use their size and stage of development to estimate the time elapsed since death.

Elsewhere in his ancient writings, Sung Tz'u reveals his own appreciation of the rapidity with which flies find a corpse in warm weather. He also advises would-be investigators to note the differences between infestations that indicate a short time since death—say, a smear of eggs or squiggles of tiny, newly hatched larvae—and those suggesting several days—as in roiling masses of chunky maggots. Whether the Chinese continued to practice forensic entomology after the thirteenth century remains unknown, as mention of it disappears from their historical accounts. In the West, by contrast, superstition obscured the usefulness of insects in time-of-death determinations for another 400 years.

* * *

EUROPE IN THE Middle Ages was a place where mud gave birth to wasps, musty hay to mice, and rotten meat to maggots. Even at the dawn of the seventeenth century, Western science remained under the sway of Aristotle, who advanced the Greek belief in spontaneous generation in the fourth century B.C. "Some creatures come into being neither from parents of the same kind nor from parents of a different kind, as with flies and the various kinds of what are called fleas," Aristotle wrote in *De Generatione Animalium* (On the Generation of Animals). "Such are all that come into being not from a union of the sexes, but from decaying earth and excrements." This idea, that insects and assorted "bloodless creatures" could spring to life fully formed, short-circuited any scientific study of their actual life cycles—the crucial first step in their use as a time marker for death.

Over the next 200 years, belief in spontaneous generation would come to permeate Western thought, as expressed by Lucretius, Rome's most distinguished scientist-poet:

> *With good reason the earth has gotten the name of mother*
> *Since all things are produced out of the earth.*
> *And many living creatures, even now, spring out of the earth*
> *Taking form by the rains and heat of the sun.*

Today such beliefs may strike us as childish. To understand how they could remain "science" for more than two millennia is to appreciate how Aristotelian thought exalted the value of pure logic over the "baser" approach of objective testing. This led to a rigid set of rules for pursuing truth through the unsullied application of the mind, or deductive reasoning—in direct contrast to modern scientific method with its emphasis on controlled experimentation. For example, if one starts with the observation "no flies rise from the banks of a river when they are dry" and adds the observation "flies do rise from the banks of the river after a rain," Aristotelian logic gives you "therefore, rain plus dry earth (that is, mud) produces flies."

What the ancients failed to appreciate was that such syllogisms work only when the "thinker" possesses *all* the facts, just as a computer readout is only as good as its data input. In deductive reasoning as in computer programming, "garbage in, garbage out." Consequently, the Greeks' superficial observations, coupled with rigid logic, produced a strangely twisted notion of reality in which snakes sprang from horsehair dropped in a bucket of milk. Admittedly, before modern refrigeration and storage methods, seeming proof of spontaneous generation appeared at every turn. Left on the table long enough, cold meat did "sprout" maggots. Mealybugs sprang from flour. Weevils, from nuts. Even a glass of ordinary water eventually grew turbid with life.

In the early 1600s, the respected Belgian physiologist Jan Baptista von Helmont went so far as to confirm such everyday observations with what he called "scientific proof," namely his own laboratory method for generating mice. Von Helmont's formula called for combining soiled and sweaty undergarments with several stalks of wheat in an open jar or pot, then waiting twenty-one days for the sweat from the rank underwear to penetrate the husks and convey the necessary *animus,* or life force, to change the grain into scurrying mice. Yet another seventeenth-century recipe for producing mice entailed leaving cheese wrapped in rags in the corner of a room.

The first significant crack in the theory of spontaneous generation came in 1667, at the hands of the celebrated Italian scientist-poet Francesco Redi. A keen-witted and dashing courtier, personal physician to the Grand Duke Ferdinand II of Tuscany, Redi epitomized the term "freethinker" and delighted his scientific-minded employer with amusing experiments debunking fusty Aristotelian principles.

Contrary to scientific lore, Redi never concerned himself with disproving the theory of spontaneous generation, per se. As a physician, he believed that many intestinal parasites sparked to life in this way. Rather, Redi became intrigued with a novel and controversial idea put forward in 1651 by the English physician and anatomist William Harvey. According to Harvey, insects developed from *oviform,* or egglike material—just like those of a chicken except too small to see. The logic of this suggestion—that small animals, like larger ones, came from living eggs, albeit invisible ones—appealed to Redi and his belief in an orderly world.

Further inspiration came from an unlikely source: classical Greek literature. One day, rereading *The Iliad,* Homer's epic ballad of war, Redi pounced on a familiar passage in which the hero Achilles begs his mother to protect from decay the corpse of his beloved friend Patroclus, newly slain by the Trojan prince Hector. "I much fear," says Achilles, "that flies will settle upon the son of Menoetius and breed worms about his wounds, so that his body, now he is dead, will be disfigured and the flesh will rot." Reassuring Achilles, his mother, Thetis, replies: "My son be not disquieted about this matter. I will find means to protect him from the swarms of noisome flies that prey on the bodies of men who have been killed in battle." Redi is said to have literally sprung to his feet at these words . . . for what use could it be to protect a body from flies, if their "worms" sprang directly from decaying flesh? Clearly Achilles and Thetis—or at least their chronicler Homer—understood on some level that flesh-eating maggots came from eggs laid by flies.

Taking his cue from Thetis's promise to keep the "noisome flies" at bay, Redi became obsessed with designing the perfect experiment to move the "maggot question" out of the arena of philosophical debate into that of empirical fact. Setting Redi's experiments apart from anything that came before them was his clear-headed use of "controls," or comparison tests, designed to be alike in every way save for the "variable" in question.

Specifically, Redi's classic investigation (which has been replicated by countless nauseated high-school students) involved placing chunks of meat in wide-mouth jars and leaving them to rot in the open air. Some jars he covered with a fine wire mesh to seal out flies. Others he left open. The famous result: Both covered and uncovered meat turned rank, but only the uncovered, or "fly-accessible," meat became maggoty.

Although this standard account of Redi's hallmark experiment conveys the gist of his conclusions, it fails to capture the utterly exhaustive lengths to which he took his proof. Not only did Redi repeat his maggot test using both cooked meat and raw, he replicated it with bull and calf, horse and colt, buffalo, ass, fallow deer, chicken, sparrow, swallow, fish, and assorted reptiles. Courtesy of the Grand Duke's personal zoo, Redi even tested the meat of lion and tiger. (Given the pitiful survival rates of zoo animals before the twentieth century, chances are Redi obtained his exotic meats as the result of malnutrition and neglect, not scientific sacrifice.)

Redi also noted how certain female flies—when prevented from reaching their goal by the wire mesh—extruded the tips of their abdomens in a desperate effort to pass their eggs inside. Redi proved correct in predicting that tiny maggots would appear on the wire mesh a few days later, though they quickly died for lack of meat and moisture. In one instance, Redi even observed a live-bearing flesh fly deposit two, tiny maggots directly on the screen. In this isolated case, the tiny larvae managed to squirm through the mesh and drop onto the meat below to continue their development.

In pioneering the "controlled experiment," Redi both elevated the scientific method to new heights and established the zero-point from which forensic entomologists would someday calculate time of death—that is, the arrival of egg-laden flies, drawn within minutes by the irresistible scent of death.

As Redi no doubt expected, his conclusions met immediate attack, not by the scientific establishment but by church officials, who accused him of impugning the authority of the Scriptures. In arguing Redi's heresy, they cited a classic passage in the Book of Judges, in which the biblical hero Samson challenges the Philistines with the riddle "Out of the eater came forth meat. And out of the strong came forth sweetness." (What is it?) The answer to Samson's riddle, as betrayed to the Philistines by his unfaithful bride, was "bees," which our hero had personally seen rising from the carcass of a lion.

Secure in his position as an adored court physician, the waggish Redi rose to the fight. Not only did he stand firm on his rejection of spontaneous generation, he gleefully vowed to go one step further and refute what he considered the Church's "tedious" comment on the vanity of man—namely the standard canon that human flesh is "mere pasture for filthy flies." Redi did so by demonstrating that all one had to do was bury a corpse quickly enough, before flies could lay their eggs, and it would remain untouched by flies and their maggots indefinitely. (Though Redi and his intellectual cronies must have gloated in this last, smug demonstration, they clearly did not push its limits with long-term studies. If they had, they would have discovered that, given enough time, certain flies could indeed burrow down into graves to lay their eggs.)

In 1668, Redi published his masterpiece: *Esperienze Intorno alla Generazione degli Insetti* (Experiences Concerning the Generation of Insects). A keen naturalist, he went on to identify four of the flies attracted by the meat's odor. In addition to the common house fly, they included three that would become of great forensic interest: the flesh fly, blue blow fly, and greenbottle. Quick to arrive on the scene of death, the husky flesh flies and their colorful cousins, the iridescent blow, or bottle, flies could be found in virtually every region of the world. The latter took their common name from their habit of inflating, or "blowing," meat with their wriggling young. The synonym "bottle," as in greenbottle, may derive from the old Gaelic diminutive for "bot," meaning maggot.

As poetic emblems of death, the flesh and blow flies already graced Renaissance literature, as when Shakespeare's Ferdinand professes that, save for the love of Miranda, he would sooner "suffer the flesh-fly blow my mouth" (*The Tempest*, act 3, scene 1), or when the Bard's Cleopatra chooses death over captivity. "Rather on Nilus mud, lay me stark naked and let the flies blow me into abhorring" (*Antony and Cleopatra,* act 5, scene 2).

Redi's famous experiments appear to have sparked similar interest among Europe's scientists, who focused their newly invented microscopes on the Lilliputian insect world to glimpse the tiny eggs that Redi predicted they would find. In 1669, the Italian anatomist Marcello Malpighi and the Dutch biologist Jan Swammerdam separately published landmark studies on insect anatomy, effectively launching the modern science of entomology. The study of insects continued to flourish in the eighteenth century, although its students had yet to grasp the magnitude of their field. Over the next century it would become clear that insects outnumbered the rest of the planet's animal life combined.

This was also the age of the scientist as collector, intent on killing and mounting at least one of virtually every living creature he or she encountered. Prominent among any credible insect collection would have been an array of large, colorful blow flies. Biologists with a more than idle interest in entomology went further, to document the life cycles of the insects in their collections. Because their maggots grew readily on a bit of meat in a well-aired jar, flies were again among the first insects studied. Their four-stage life cycle—egg, larva, pupa, and adult—became the standard against which all other insect life cycles would be measured. It would likewise become the basis of the fly's usefulness as a postmortem stopwatch.

With new species being discovered at every turn, early fly experts, or dipterists (from the Greek *dipteros,* "having two wings"), largely confined their work to identification, anatomical description, and the clarification of basic life stages. Yet the mere act of focusing attention on the ubiquitous flies led to an appreciation of their powerful role in decomposition. The first hint of this discovery appears in the writings of the great Swedish naturalist Carolus Linnaeus. Although his journal entries reveal a fastidious man easily nauseated at the sight of "raw fish with mouths full of worms," Linnaeus's drive to classify "all God's creatures" included what he termed "life's vilest." In creating his historic classification system in 1757, Linnaeus lumped blow and flesh flies together with the common house fly in the family *Musca.* In 1767, he mentions them again, almost admiringly, writing "tres muscae consummunt cadaver equi aeque cito ac leo," or "the progeny of three flies can consume a dead horse more quickly than can a lion." A decade later, England's Moses Harris would take a close look at the delicate venation of fly wings and use it to divide Linnaeus's *Musca* into five families, including what would become the modern-day Calliphoridae, or blow fly family, and the Sarcophagidae, or flesh flies.

Just as important, the orderly allure of scientific classification helped fuel the study of insects as a fashionable pastime. Investigations into *res naturae,* "the things of nature," became immensely popular among the educated European gentry in the late eighteenth and early nineteenth centuries. Pleasant Sunday afternoons found men and women of leisure hiking through woods and pasture, overturning rotting logs and other dead matter to observe the furious activity beneath. Their amateur investigations filled early scientific journals and helped broaden and deepen knowledge of insect behavior.

Against this backdrop, forensic entomology made its first appearance in a Western court. The precedent-setting case traces to an early spring day in 1850 at the boardinghouse of Mademoiselle Saillard, in the French town of Arbois, in Burgundy. Repairing a plaster mantel piece in one of Saillard's apartments, a workman accidentally broke through the wall to the chimney behind it, immediately releasing a sharp odor. Tracing the smell to its source, the horrified plasterer found the mummified remains of a newborn baby. The workman's fright, however, paled in comparison to that of the young couple living in the flat. Despite their claims of innocence, they stood accused of murder, or at least an illegal abortion.

As luck would have it, however, the medical examiner summoned to the scene—Dr. Marcel Bergeret—fancied himself a naturalist as well as a medical doctor. In summarizing his approach to the crime, Bergeret jotted down five questions he would have to confront. First, was the child born at term? Second, was it alive when it was born? Third, how long did it live? Fourth, how did it die? And fifth, what was the time interval between death and discovery?

Bergeret's physical examination revealed that the child had been born or aborted prematurely: Its bones had not fully formed. Establishing a date for its stillbirth or brief life proved far more challenging. The tiny body's advanced mummification offered few clues. The extremely warm, dry atmosphere inside the chimney had quickly shriveled its delicate tissues. Examination of the organs likewise seemed fruitless, because insects had already devoured them.

Yet any criminal proceedings hinged on determining the time of death. Indeed, four different sets of tenants had lived in the flat during the previous three years, and the local magistrate depended on Bergeret's examination to determine which tenants would stand accused of the crime. It was then, Bergeret says, that he experienced a moment of "unexpected enlightenment." Specifically, Bergeret realized he might be able to construct a time line using the insects he'd collected from the body. He writes,

> The eggs of the larvae we found on the corpse in March 1850 must have been deposited there in the middle of 1849. Therefore, the corpse must have been deposited before this time interval. Next to the many living larvae there were numerous pupae present, and they must come from eggs that have been laid earlier, i.e., in 1848. . . . Could it be that the corpse was deposited even before that time? The fly that emerges from the pupae that we found in the body cavities, is *[Sarcophaga] carnaria* [a flesh fly], that lays its eggs before the body dries out. The larvae were of little butterflies of the night [moths] that attack bodies that are already dried out. If the body was deposited, say, in 1846, or 1847, we would not have found those larvae [since they would have hatched]. In conclusion, two generations of insects were found on the corpse, representing two years postmortem: on the fresh corpse, the flesh fly deposited its eggs in 1848, on the dried out corpse, the moths laid their eggs in 1849.

Bergeret clearly erred in assuming that the insects in question would have each taken a full year to complete their life cycle. In warm weather, they can do so in a matter of days to weeks. Yet his reasoning follows two of the founding axioms of what would become the science of forensic entomology: First, that certain insects feed exclusively on fresh remains (in this case, the flesh flies) and that others wait until the flesh dries (the moths). And second, that insects require warmth to be active—hence, Bergeret's conclusion that both the flesh flies and the moths would have found the body in summer.

Duly impressed with Bergeret's elegant bracketing of the baby's time of death, the local French magistrate dismissed charges against the tenants then living in the flat and issued arrest warrants for the couple who lived there two years earlier.

To this day, Bergeret's clear commentary on the successive colonization of a corpse—with flesh flies colonizing the fresh body and clothes moths attracted to the dry remains—stands out as a simple, yet sterling example of the tricky art of determining time of death with "faunal progression." The term refers to the more-or-less orderly waves of different insects, spiders, mites, and other arthropods that wash over an aging corpse in the weeks, months, and years after death.

Publicity surrounding the Arbois baby murder and Bergeret's account of the case in a French medical journal sparked interest in the forensic value of insects and their arachnid cousins, spiders and mites. Not surprisingly, the keenest interest came from forensic pathologists stymied by their inability to determine time of death by conventional means. French pathologists, in particular, exalted the Bergeret case to the level of medical folklore. Yet from written records, it would appear that no European medical examiner took the logical next step—to involve a proficient entomologist in a postmortem study—for another quarter century.

This crucial bridge was finally built in 1878, shortly after a Parisian magistrate sent the body of a newborn child to the morgue of the highly esteemed forensic pathologist Paul Camille Hippolyte Brouardel. The tiny body had been found in the underbrush of an abandoned lot on Rue Rochebrune, an otherwise bustling residential boulevard running through Paris. The sight of the hollowed-out, half-mummified infant immediately brought to Brouardel's mind the famed Arbois baby case. In addition, he noted swarms of mites moving through a stringy, brown powder that coated the cadaver's skin and body cavities.

Hoping to replicate Bergeret's clever time-of-death determination, but with no pretensions of insect expertise, the master coroner visited the offices of Professor Edmond Perrier at the nearby Museum of Natural History, where he was directed to the laboratory of Jean Pierre Mégnin, a veterinarian and respected researcher specializing in parasites, in particular the eight-legged kind. For many years, mites had been the objects of Mégnin's favorite studies.

Following Brouardel back to the morgue, Mégnin identified the stringy, brown powder on the body as "frass"—the excrement of skin beetles that had no doubt fed on the dried remains before abandoning the body the instant the police disturbed it. The more tenacious mites Mégnin identified as belonging to the family Acaridae, a large group of minute arachnids known best for their habit of infesting preserved meat and animal hides. From these findings, Mégnin could offer only a rough approximation, that the baby had died at least six months, perhaps a year, earlier—time enough for the tissues to dry and become attractive to the gnawing invertebrates.

More important for the future of forensic entomology, Mégnin found himself utterly transfixed by the lurid yet compelling science practiced at the Paris morgue. As Brouardel and his colleagues shared their frustration in determining time of death with accuracy, Mégnin came to understand the vast potential for using insects as forensic clues . . . if only more were known about their habits and life cycles. In the months and years that followed, Mégnin became a regular visitor to the morgue's damp, underground examining rooms. He also accompanied Brouardel and the other medical examiners to the more degenerate parts of the city on homicide investigations.

The Parisian pathologists welcomed Mégnin's interest and input, and sent him word whenever a "well-colonized" corpse passed through their venue. They also put in the entomologist's hands the works of the forensic chemist Matthieu Orfila, the venerated father of modern toxicology—the study of poisons. Orfila, in developing postmortem tests for detecting arsenic and other death potions of the day, was among the first scientists to systematically record bodily decay in exhaustive detail. In his 1831 treatise on juridicial exhumations, Orfila noted the same species of flies as Redi had in his classic experiments, and he echoed Redi in his description of the dramatic way maggots could alter the decomposition process, depending on whether flies had access to a body before burial:

> What if one now buries two cadavers, of which the one offers at its surface thousands of eggs, whereas the other hasn't yet any? It be-

> comes evident that the first will putrefy much more rapidly, all other circumstances being equal. This is because it is the nature of the larvae to destroy our tissues in order to nourish themselves. One then must not fail to notice the role of insects on body surfaces in the putrefaction process.

From his own observations, Mégnin recognized that Orfila's conclusions were basically correct but riddled with inaccuracies owing to the toxicologist's ignorance of blow fly behavior. In particular, Orfila asserted that only bodies left exposed till they putrefied, or stank, attracted the attention of flies, writing, "It is proved that in the first period after death, the flies don't settle around the cadavers and that later they only fly about near them, and that finally when decomposition is greatly advanced, they alight on them and deposit their eggs."

Mégnin understood that in warm weather, blow flies typically found man or beast within moments of death, at times even before the heart had stopped beating. In particular, Mégnin noted the marked attraction blow flies show for the body's "natural openings." He concluded:

> From these places, certain emanations issue out to signal to [the blow flies] the imminence of an event that will soon procure an abundance of food for their progeny. As a result, these flies already fall furiously upon [the body] attempting to lay eggs in the nostrils, in the mouth, or even in the eyes.

With this observation came Mégnin's conviction that he could forge out of the blow fly a postmortem clock more accurate than anyone had imagined—an indicator for pinpointing time of death not just to a particular year or month, but to a week or day, perhaps even to a range of hours or minutes. Mégnin likewise began recording the arrival and departure of other "squads" of insects, finding that they populated a corpse in predictable waves. "We have been struck by the fact that we have been the first to observe that the insects of cadavers, the workers of death, only arrive at their table successively, and always in the same order," he wrote.

In attempting to nail down the parameters of each wave, Mégnin envisioned constructing a reliable, month-to-month, year-by-year calendar of necrophilous, or "dead-loving," insects. So inspired, Mégnin all but abandoned his research on mite infestation, though at the time he stood at the pinnacle of his career as a veterinary entomologist. It would be the beginning of nearly three decades of exhaustive study on the uncharted *la faune des tombeaux,* "the fauna of tombs."

In addition to his observations at homicide scenes and the Paris morgue, Mégnin began frequenting local cemeteries, persuading caretakers there to allow him to catalog insects on bodies being exhumed during plot repairs and renovations. Mégnin's longtime colleagues at the University of Paris and the Museum of Natural History were aghast at the bizarre transformation in the professor's scientific interests. Their disdain stemmed not so much from the macabre nature of Mégnin's studies, though work with the dead remained highly controversial, but from Mégnin's deviation from pure research. Had Mégnin restricted himself to dry, ecological studies of necrophilous insects, he likely would have remained within the realm of respectable academic inquiry. What sullied Mégnin's reputation among the academic elite was his new habit of rubbing elbows with police detectives and involving himself in their investigations. Mégnin had lowered himself out of academia's ivory tower and into the vulgar realm of the secular, no less than if he had become a bug exterminator.

Ignoring such criticism, Mégnin published his *La faune des cadavres: Application de l'entomologie à la médecine légale* in 1894. Reports of the book's gruesome subject matter captivated the public, and the term "forensic entomology" pierced Western thought for the first time. The publication's reception in scientific circles remained mixed. Not that academics faulted Mégnin's meticulous and exhaustive research; the problem lay in the sensational publicity it generated. The unseemly public spectacle only deepened contempt for Mégnin's work among so-called serious biologists.

By contrast, forensic entomology became a cause célèbre among medical examiners and forensic pathologists, already inured to the gruesome nature of their field. They hailed Mégnin's work as delivering forensic science's long-awaited holy grail: death's own timepiece. Specifically, Mégnin had documented eight distinct waves of insects and other arthropods that washed over any corpse exposed to the open air. "It is the flies who open the march of Death's Workers," he wrote. Quickest of all, he added, were the blow flies, or bottle flies, which arrived in the first minutes after death or grievous injury. Flesh flies followed within a day or two, drawn by the smell of early decomposition.

Mégnin's third through fifth waves crested and ebbed between the third and eighth month after death and comprised a variety of rove beetles and soldier flies that preyed on the developing maggots, as well as cheese skippers and scuttle flies that fed on the fermented protein products of the later stages of decomposition. As the last bodily fluids disappeared in months six through twelve, mites swarmed over the dry cadaver, followed by skin and hide beetles in the second year. Wrapping up the show, three years after death, clothes moths and spider beetles scavenged the completely desiccated remains. Magnin concluded:

> In the end, nothing rests next to the white bones but a sort of brown earth, finely granular, composed of insect pupal cases . . . and the excrements of successive generations of insects . . . thus is accomplished this parable of the scripture: You are dust and unto dust you shall return.

Mégnin documented a somewhat simpler and more drawn-out parade of insects on *buried* corpses. The progression usually began with blow flies, although their access to the body could be diminished somewhat by weather or speedy burial. Of great importance to later homicide investigations, Mégnin noted that only cadavers buried in winter remained consistently free of blow fly infestation. "In short," he wrote, "when one proceeds in burying a dead person during the summer, one closes up many wolves in the sheep fold."

Once buried with their repast, blow flies often succeeded in producing more than a half-dozen generations before ultimately dying off inside the casket, Mégnin claimed. Indeed, over the course of the first year, they "occupy the workshop of death alone." Only after eight to twelve months following burial did the odor of rancid fats and ammonia percolate up through the soil to lure other visitors: The second wave consisted of a few sturdy dump flies; the third wave, swarms of gnatlike coffin flies. Both types of fly laid their eggs on the ground above the coffin, their larvae able to burrow deep into the soil and through the cracks of the best-sealed casket. A fourth wave of root-eating beetles wrapped up the underground procession in the second and third year.

Soon after its publication, *La faune des cadavres* became a frequent reference in European homicide cases. Indeed, the 214-page volume, with twenty-eight illustrations to aid insect identification, could be found in the offices of medical examiners and scientifically minded police detectives throughout Europe and North America. So armed, many eager young Mégnin fans attempted their own entomological determinations, a trend that clearly alarmed the master. Mégnin argued that a decade and a half of field studies by one man could do no more than lay a foundation for a new field of scientific inquiry. As an entomologist, Mégnin knew that insect species varied greatly from one geographic area to another. He questioned whether his observations held true outside of Paris, let alone France. He also raised questions as to how variables such as weather, soil type, and insect predators might alter the progression of "death's faithful workers."

Urging further entomological research, Mégnin took issue with individuals who misinterpreted findings as more precise than they were. Others concurred. In 1897, an article in the *Montreal Medical Journal* cautioned North American pathologists against embracing the new fad of "medicocriminal entomology" too quickly. "The chief danger to be feared from Mégnin's imitators," the authors wrote, "is that they may indulge in guesses without solid basis to apply rules to countries and climates where they are inapplicable." Hoping to rectify the situation, Canadian pathologists Wyatt Johnston and Geoffrey Villeneuve cataloged insects on several bodies that had been exposed to open air, and published their findings.

But the task ahead was gargantuan. At the end of the nineteenth century, the insect fauna of the United States and Canada remained virtually uncharted. Even by modest estimates, there were likely hundreds, if not thousands, of different insects and other arthropods that might be found on corpses in different regions and habitats. Although only a fraction of these species were known to science in 1898, fewer still had been studied with the aim of documenting the time span of their development from egg, through larval stages, to adult. And none had been studied as to its preference in decaying flesh—that is, fresh, bloated, or dry. Complicating the picture further, entomologists knew that variations in conditions such as temperature, sunlight, and humidity could skew insect arrival and development.

Even in Europe, with its centuries-old tradition of supporting academic research, such detailed understanding of noneconomically important insects remained scant. In the New World, it was virtually nonexistent. North America lacked Europe's tradition of giving its scholars free rein to pursue research of their choosing, however idiosyncratic, abstract, or impractical. The U.S. government expected entomologists in its employ to address pressing needs such as the control of agricultural pests and insects involved in the transmission of disease. Entomologists employed by private universities, in turn, remained swamped with teaching duties.

Meanwhile, in Europe, isolated interest in studying "death's fauna" kept entomology's forensic potential on a low simmer. In the more classical style of European research, several German entomologists continued to catalog insects found in graves and on decaying animal carcasses, steering clear of any actual forensic cases. In fact, it took one of England's most sensational murders to revive interest in forensic circles.

* * *

ON THE MORNING of September 29, 1935, a young woman crossing a bridge near Moffat, Scotland, glanced down to see over a dozen packages littering the stream bank below, some bound in newspaper, others in strips of cloth. Her curiosity piqued, she looked closer, only to realize with horror that one partially ripped parcel contained a human arm. Police from nearby Edinburgh arrived that afternoon. Their search turned up a gruesome collection of butchered body parts. In all, some seventy maggot-infested fragments were found strewn through the ravine, which has ever since been known as the Devil's Beef Tub.

Despite the profusion of body parts, including two heads and two partially smashed pelvises, the police remained clueless as to the victims' identities. Whoever disposed of the bodies had meticulously chopped off or gouged out all identifying features—not only eyes, ears, and other facial features, but even fingertips and sexual organs. Reassembling the skeletons as best they could, pathologists at the University of Glasgow concluded they were dealing with one female and one male, as one set of bones appeared notably larger and sturdier than the other. They likewise estimated that the couple had been killed on or just before September 19, with their dismembered parts dumped into the Moffat stream soon after. This concurred exactly with the area's last heavy rain, which would have swelled the stream to levels high enough to widely scatter the murderer's grisly parcels.

By chance, one of the Scottish constables called to investigate the double murder had the previous day read a short newspaper account of a disappearance 100 miles to the south, in Lancaster, England. The wife of a prominent local doctor, Isabella Ruxton was last seen alive on September 14, after which she vanished, along with the family nursemaid, Mary Rogerson. Though the missing person report didn't match the pathologists' profile of a male and female couple, a report of two persons disappearing together seemed unusual enough to pursue.

Mrs. Ruxton's neighbors had come to the Lancaster police the day after her disappearance, fearing the worst. The missing woman's husband, Dr. Buck Ruxton, had been making a particular spectacle of himself of late, raging on about his wife's alleged infidelities to anyone who would listen. Friends knew of the doctor's often violent fits of jealousy. They also described the family nursemaid as a particularly robust, "large-boned" woman.

When questioned, Dr. Ruxton protested that the two women had merely gone on holiday to Edinburgh the day of their so-called disappearance. He even angrily demanded that police search his home to quash the vicious rumors that were circulating around town. Dr. Ruxton's bluff proved unwise: police found traces of blood in the carpeting and fat, possibly human fat, in the drains of the bathroom sink. Meanwhile, north of the border, investigators traced a scrap of newspaper from one bloody parcel to an edition sold only in Lancaster.

Police arrested Dr. Ruxton on October 12. However, the case against him remained shaky. In particular, investigators worried that the medical examiner's initial time-of-death estimate—on or just before September 19—made plausible Ruxton's defense, namely that the women must have met mischief while on holiday in Edinburgh. Indeed, Ruxton had witnesses to vouch for his whereabouts in Lancaster for not only September 19, but also the day before and after.

The constable in charge of the case went back to the pathologists who had pieced together the bodies, asking if they might have erred, if the murders might have taken place earlier. "Certainly possible," they replied. But medical science lacked any concrete indicator to warrant a change in the initial estimate. At this point, one of the doctors pulled out a vial of fat maggots he had removed from the mutilated remains. In a move worthy of Scotland Yard lore, he suggested taking the specimens down the hall to the laboratory of entomologist Alexander Mearns.

Mearns identified the larvae as that of the common bluebottle fly, *Calliphora vicina*. Further, he found the oldest of the larvae to be at the end of their third instar, or life stage. Assuming the bodies had been dumped in the obscurity of night or early morning, Mearns drew up the following timetable: The first flies would have arrived six to ten hours after the dumping, in the warmth of late morning. From then, it would take another eight to fourteen hours for the eggs to hatch, and an equal amount of time for the tiny, first instars to complete their development. Maturation would then slow as the second-instar larvae fattened over two to three days, with another seven to eight days for the maggots to complete their third stage of development. Since the weather had been chilly, Mearns favored the longer end of this time span, approximately twelve days before the remains were examined—and the insects were collected—on October 1. His conclusions perfectly fit the prosecution's assertion that Ruxton had murdered the women on September 15, dumping their packaged remains before morning light on September 16.

Together with corroborating evidence, the entomologist's testimony led to Ruxton's conviction. Following the doctor's execution in 1936, newspapers published a jail-cell confession in which he admitted strangling his wife in a jealous rage, then killing the nursemaid when she stumbled on the scene. The sensational case became a criminology classic when the forensic team at

Glasgow University detailed their findings in a book, *Medico-Legal Aspects of the Ruxton Case* (1937).

The Devil's Beef Tub murders marked the beginning of an era of scientific crime solving when homicide investigators would reach out to nonpolice experts who might shed light on a curious artifact or a seemingly out-of-place object—be it a squashed berry on the sole of a suspect's shoe or a stray fiber beneath a victim's fingernail. If anything, it was the sheer ubiquity of insects on decomposing bodies that inured police to their presence, and prevented them from consulting entomologists more often.

Still, on rare occasion a detective—invariably hard up for clues—would show up at a natural history museum or university with a vial of wiggling, white maggots or a handful of dark, pellet-like puparia, asking if an entomologist could identify them and offer an opinion as to their age. Dipterists at such institutions in the 1930s related anecdotes of such visits, but the visits remained no more than brief diversions from the routine. "It was all very hit and miss in those days," relates Kenneth Smith, curator emeritus of the British Museum in London. "The constables would show up, get a crude time estimate based on what the curators could find about a particular insect in the [scientific] literature and disappear again."

Presumably, such clues helped guide police work at times, perhaps by narrowing a field of suspects or suggesting new directions for investigation. However, prewar records show no case but the Ruxton murders in which an entomologist actually took the witness stand. Moreover, no scientist of the day would have considered him- or herself a forensic entomologist in the tradition of Mégnin. In fact, familiarity with Mégnin's research had all but disappeared by the 1950s, when an obscure murder in the Eastern Bloc would remind investigators that entomology could, at times, pinpoint death to as close as a few hours.

The investigation began on a chilly September night in 1956, when police found the knifed body of a mail carrier shoved behind some pilings near a local ferry dock. With no apparent motive for the killing, police began questioning anyone who might have been on the scene that evening. Their suspicions settled on a ferry skipper who had reported to work at 6 P.M., just a few hours before the body's discovery. Although accounts of the investigation remain sketchy, it appears a knife of suitable size to have made the deadly wounds was found either on the skipper or somewhere between his dock and the murder scene. Ignoring the accused's predictable claims of innocence, the police arrested him and held him for trial.

At 4 P.M. the day after the body's discovery, the local coroner performed a thorough autopsy. In his report, he noted masses of yellowish fly eggs and several squirming larvae about one to two millimeters (.04 to .08 inches) in length. But neither prosecution nor defense took note of the coroner's footnote at trial. The presiding judge sentenced the skipper to life in prison.

Imprisonment did not diminish the convicted captain's passionate claims of innocence. Eight years later, he succeeded in getting his case reopened. This time, the defense called to the stand Ferenc Mihalyi, curator of entomology at the Budapest Natural History Museum. The captain could not possibly be guilty, Mihalyi asserted. Why? Because it was known that the convicted man had not arrived in the vicinity of the pier until 6 P.M. The month being September, that would have meant the sun had already begun to set—making it too dark and chilly for any blow fly to be active.

Mihalyi further testified that he himself had performed experiments with the yellowish eggs of the greenbottle fly, *Lucilia caesar.* In these studies, the eggs had taken a full thirteen hours to hatch at a warm laboratory temperature of 26 degrees Celsius (78 degrees Fahrenheit). In the cold night air and then the air-conditioned chill of the morgue, the eggs would have developed more slowly. Mihalyi had likewise studied the other two species of blow fly native to Hungary—the sheep blow fly, *Phaenicia sericata,* and the Holarctic blow fly, *Protophormia terraenovae.* Their eggs required ten to eleven hours and fourteen to sixteen hours to hatch, respectively.

So it was simply not possible that the larvae found on the postman's body could have come from eggs laid in the morgue on the day of the autopsy. They must have been laid the previous day during daylight hours, when witnesses could vouch for the skipper's whereabouts. The judge, intrigued but skeptical, ordered Mihalyi's experimental data to be verified. They were, and the captain walked free. Pursuing other leads, the police eventually found the real killer. As it turned out, his motive had been to steal the paychecks that the mail carrier normally carried on the tenth day of each month. By chance, the checks had been delayed, and so no theft had been noted.

Despite the unprecedented accuracy of his forensic determination, Mihalyi neither published a scientific account of the case nor testified again in a court of law, but quietly returned to his studies of the seasonal distribution of Hungarian flies. Ironically, it would be in America—where forensic entomology had lain forgotten for over a half century—that the science would finally mature beyond occasional dalliance into a recognized field of study.

5 Bug Sleuthing Crosses the Atlantic

America is a land of wonders, in which everything is in constant motion and every change seems an improvement.

—ALEXIS DE TOCQUEVILLE (1805–1859), *DEMOCRACY IN AMERICA*

ON A CRISP fall evening in 1829, Harvard librarian Thaddeus William Harris sank into the chair behind his writing desk, lit a small lamp, and adjusted its wick to a bright, clean burn. Pulling out paper and pen, Harris folded back his starched white cuffs before dipping the stylus. After a long, lackluster day shelving books, balancing accounts, and directing hapless students, he was looking forward to losing himself in "conversation" with his good friend and colleague in science, Edward Doubleday, in England.

Like Harris, Doubleday was an entomologist by training and passion. Their correspondence overflowed with earnest debate over new manuscripts and questionable identifications. Occasionally Doubleday described a pleasant weekend he'd spent collecting specimens in the English countryside, assisted by one or more admiring young ladies skilled in biological drawing. Doubleday confided how he and his companions envied Harris his vast New World of undiscovered species. Like many Europeans, they took a lively interest in the "insect exotica" of North America, a fascination that dated to 1587, when John White, commander of Sir Walter Raleigh's third Virginia expedition, returned with a line drawing of a North American swallowtail butterfly.

By stark contrast, few Americans had ever developed an interest in such "foppish foolery." Hardened by the immediate challenges of their new frontier, early settlers generally saw bug collecting, taxonomy, and insect illustration as pastimes of the idle rich or the extremely eccentric. Even with the founding of respected science programs such as Harvard's, would-be entomologists suffered from lingering prejudice.

So although Harris's research on New England insects had won considerable respect in Europe, it couldn't earn him the slenderest income in the United States. On the contrary, his interests brought little more than derision. More than once, Harris had pulled up stone-faced when laughing townsfolk came upon him scrambling over a cemetery wall in pursuit of some flying *Buprestis* or stripping the bark from a tree in search of a shy *Curculio* larva. Far from impressing the ladies, Harris's early attempts to explain his studies had quickly earned him a reputation as a highly questionable dinner guest.

Putting pen to paper that September evening in 1827, he lamented to Doubleday,

> You have never, and can never know what it is to be alone in your pursuits, to want the sympathy, the aide and counsel of kindred spirits. You are not compelled to pursue [this] science as it were by stealth, and to feel all the time, while so employed, that you are exposing yourself, if discovered, to the ridicule perhaps, at least to the contempt, of those who cannot perceive in such pursuits any practical and useful results.

Though the then-unimagined field of forensic entomology was destined to achieve its greatest potential in North America, in the mid-1800s Harris rightly saw the continent as an entomological no-man's-land. Admittedly, even the combined efforts of Europe's tenured entomology professors and weekend enthusiasts would have been no match for the chaos of North America's insect fauna in the nineteenth century. The challenge went further than simply describing North America's native insects. Every month brought new introductions, not just from Europe, but also the Orient, Africa, West Indies, and South America. They came hidden in shipments of grain, caskets of tea, barrels of rice, cargo holds of cattle and sheep, even the hair and clothing of sailors and slave—a global confusion of flies, cockroaches, beetles, moths, scorpions, lice, fleas, mites, and more. Although many of these immigrants died off quickly enough, others thrived in an environment free of age-old predators. Complicating things further, many Old World species closely resembled their New World kin in physical form, if not in habits and life cycle.

For good reason, in 1895, Canadian pathologists Johnston and Villeneuve warned their North American colleagues against trying to replicate the time-of-death determinations of France's Mégnin. The American counterparts to Mégnin's eight waves of cadaverous insects remained largely unidentified and unstudied. In 1898, retired medical doctor and amateur entomologist Murray Galt Motter attempted to collect and catalog insects in the cemeteries of Washington, D.C. But his years of effort collecting insect specimens from 150 corpses (exhumed for routine reasons such as flooding and urban development) raised more questions than answers. One of Motter's stranger conclusions was that a cadaver's race influenced the insect-assisted process of decay, as did "the mode of death, whether quiet and peaceful or violent and painful." In the end, Motter failed to identify the vast majority of his specimens down to the species. His hopes that others would do so disintegrated into dust along with his collections in the backrooms of the Smithsonian.

Yet in the background, North America's insects were demanding the attention of no less a power than the U.S. Congress. In 1853, Congress created the U.S. Bureau of Agriculture (later USDA), in large part to stem insect-related crop damage that threatened to bring the budding agricultural nation to its knees. Admittedly, the duties of the bureau's first and only bug expert, Townsend Glover, also included collecting and disseminating information on seeds, fruits, birds, and textiles. In the words of contemporary entomologist B. D. Walsh, it was "a good deal like hiring a single cradler to harvest a thousand acres of wheat, and then expecting him, in addition to cut and fetch in wood, peel and wash the potatoes, and be always on hand ready to wait on the good woman of the house."

Fortunately, the federal government recognized the bureau's inadequacy within a decade. With the Morrill Land Grant Act of 1862, it funded the opening of agricultural colleges in every state and the education of a new breed of "economic entomologists" devoted to waging war on insect pests. In the coming years, the insects themselves lobbied for greater recognition. So enormous were the Rocky Mountain locust plagues of 1874 and 1876 that western trains literally ground to a halt on their crushed remains. In response, the U.S. government appropriated the then-colossal sum of $18,000 to establish the U.S. Entomological Commission, dedicated to pioneering the use of pesticides and biological controls.

Taxonomy took off in 1881, when the USDA created the Systematic Entomology Laboratory at the National Museum in Washington, D.C. The lab's primary duty was to identify and study insects sent to it from farmers, ranchers, and agricultural agents across the nation. As the world rolled into the twentieth century, American entomologists stood ready to wage warfare on the insect hordes, but there remained much to learn about the enemy.

Fortunately for American forensics, some of the insects most useful in death investigations had cousins so pestilent as to place them high on the USDA's "Most Wanted" list. Primary among them were several families of common but little studied flesh-eating flies wreaking havoc on livestock from Maine to California. A legion of specialists would be needed to sort out the confusion of outwardly similar flies before any of them could be considered a reliable postmortem clock. On the brink of graduation from the Cleveland School of Art, class of 1922, a would-be portrait painter with a dreamy resemblance to Clark Gable would become a crucial if unlikely part of the effort.

* * *

TRUDGING INTO THE corner bodega on a crisp spring morning, David Hall plunked a nickel on the counter as he'd done every day for weeks on end and scooped up his copy of the *Cleveland Dealer.* Flipping past headlines trumpeting Annie Oakley's newest feats and the Teapot Dome Scandal, the frustrated young art student headed directly to the want ads. As usual, the search felt futile. The only employment opportunities he'd ever seen for painters involved houses or boats. Besides, as his lovely but ever-practical fiancée had recently pointed out, Hall's prospects as a portrait artist remained hampered by the fact that his paintings didn't, well, exactly resemble their subjects. Hall did far better with landscapes, but could hardly be expected to support a family with such a limited skill. Before the sweet Pauline would agree to be his wife, Hall had to find a steady job.

Hall tossed the paper into a wastebasket as he shuffled up the school steps. Disheartened, he turned to head down the corridor to morning classes when a notice on the bulletin board caught his eye. "Wanted: Scientific Illustrator" . . . a recruiter from the Department of Entomology, Ohio State University, Columbus, would be at the school that afternoon. Hall paused. Entomology . . . didn't that have something to do with word derivations?

That summer, the newly engaged David Hall found himself hunched over a laboratory bench, peering down the bright tunnel of a low-power, dissecting microscope for the first time. Having grown up on a farm, Hall thought he knew flies backward and forward, especially that biting nuisance, the horse fly—which is what lay beneath the lens of his scope.

Expecting to be mildly repulsed, he instead found himself dazzled by its beauty. Focusing on the bulging emerald eyes, he tried in vain to focus on one jeweled facet at a time. Shifting his gaze, he surveyed the delicate pattern of veins lacing across the smoky wings held rooflike over the fly's what-was-it-called, yes, abdomen, which was behind the, the, the . . . thorax. Hall found the pair of tiny knobs—no "halterers"—that he knew to look for just behind the wings. From his hurried reading of an entomology textbook, Hall knew these buttonlike balancing organs to be something unique to flies, vestiges of the second set of wings possessed by most other insects. He took a moment to marvel at how much he'd learned in so short a time.

From somewhere in the distance, Hall half heard the voice of the entomologist bending over his shoulder, explaining how to use the grid marks beneath the specimen to guide an anatomically correct reconstruction. Hall jumped, as with a metallic flash, the professor's tweezers loomed suddenly huge and bright under the microscope's spotlight. Turning the horse fly this way and that beneath Hall's lens, the instructor emphasized the importance of capturing every detail, no matter how small, for each and every particular could prove crucial in distinguishing one species from another.

Hall learned how biologists used such details as guideposts when navigating through the complexities of a taxonomic key, or species identification index. Each step in the long, keying-down process involved a checklist of details:

- Distinct head, thorax, and abdomen with six legs, antennae, jaws, and mouth; wings, when present, attached to thorax; eyes simple or compound—class Insecta.
- Abdomen without terminal tails or forceps; tarsus four- or five-segmented; abdomen not constricted and hinged; single pair of membranous wings; hind wings reduced to small knobbed structures—order Diptera—true flies.

Detail by painstaking detail, a biologist worked down through ever more specific levels of taxonomic relationships—class, order, family, genera—until he or she found the exact species to which a specimen belonged. Even greater was the challenge of creating a key for a newly found species, which was the case with some of the specimens Hall would be illustrating. Such a precedent-setting project must be approached with utter accuracy and painstaking detail.

- Tarsus having three whitish pulvillar pads, the middle one dorsal of the lateral pulvilli; wing with branches not crowded to front margin, but diverging to form a triangular cell embracing apex of wing close to front margin—family Tabandidae, horse and deer flies . . .

As the professor's voice droned on in Hall's ear, the young illustrator began to sketch the stout, broad-headed fly—head, thorax, and abdomen, triangular wings and jointed legs, going back to add each groove and pit, scale and spur, spine and hair on its stunningly complex body.

All seemed to be going well through Hall's first weeks on the job. But he and Pauline had barely begun making wedding plans when trouble began. A disgruntled faculty member stood waiting at Hall's lab bench one morning, jabbing a finger at the drawing the eager apprentice had left for him late the night before. It was the consummation of a week's worth of eyestrain and headache-inducing concentration. "What's this?" the professor demanded, pointing to a pair of bristles drawn on the fly's thorax. "They're not supposed to be there."

Admittedly, Hall didn't know what was *supposed* to be where. He drew what he *saw,* or so he thought.

Chastened, the lowly technician took his masterpiece back from the grumbling scientist and promised an immediate fix. This was not good. Hall could not afford to fall behind schedule. His artwork was crucial to the department's forthcoming bulletin on the horse and deer flies of North America.

With its stealthy flight and slicing bite, the female members of this group had long been a plague on cattle, horses, and mules, as well as a painful nuisance to those who worked around them. Because the horse fly's saliva contained an anticoagulant that prevented clotting, blood loss from unchecked bites could bring a horse or other animal to its knees. An effective eradication campaign relied on entomologists being able to physically recognize and distinguish between dozens, possibly scores, of similar species, since each unique life cycle might offer different windows of opportunity for pest control.

Hall's role as the horse fly bulletin's illustrator, though technical in nature, would be crucial to its success. North American entomologists were only then beginning to sort out their continent's profusion of flies, so the scientific publication stood to be a major, even historic, contribution to the field . . . or not.

Retrieving the pinned horse fly that had been his model, Hall returned to his wooden stool at the back of the laboratory, switched on the dissecting scope atop the black marble bench, and positioned the specimen over the grid beneath the lens. Focusing sharply, he looked again. There they were, two hairs, just as he had drawn them. Gently turning the specimen, Hall noted an identical pair on the other side. He looked at his drawing. He looked back through the microscope. He looked back at his drawing. Right place, right size, right position.

Over the coming months, similar scenes played out again and again. Faculty and graduate students complained that Hall was moving bristles and spines from their proper positions, drawing structures not in their keys, including anatomical details they'd never seen before. Time and again, Hall patiently but firmly pointed out the minute appendages and markings under the lens of his dissecting scope. Sometimes the entomologists left quietly. A few offered a hearty pat on the back. Either way, Hall enjoyed the greatest discovery of all. He was good at this, very good.

With the successful publication of the department's horse fly taxonomy, Hall's workload eased. So it was with some trepidation that he approached the group of faculty members huddled around his lab bench one afternoon. Though none of them could deny his talent as an illustrator, Hall realized that his unbending attitude had rankled. Would the lull in work be the faculty's excuse to get rid of him?

What the professors wanted, it turned out, was for their young illustrator to get a "proper" education. Were his drawings somehow lacking? he asked. Not at all, the group reassured. But with his talent, his obvious interest, his keen powers of observation, he should be an entomologist in his own right. They would gladly continue to supply work with enough flexibility to allow him to pursue a full schedule of classes toward an agriculture degree, with a major in their own department of entomology.

Ten years later, David G. Hall would be called to Washington, D.C., to fill the position of Curator of Diptera at the National Museum of Natural History—becoming, in essence, the U.S. "Fly Man in Chief." In the midst of the 1930s, despite the terrible deficits of the Great Depression, Congress fully funded Hall's position, which he dedicated to research on the insect families Sarcophagidae and Calliphoridae—the flesh flies and blow flies that were then known primarily as livestock pests. As fate would have it, these were precisely the insects that would someday step forward as some of North America's most important murder witnesses. Hall recognized this arcane potential at the time, but had little interest and less time to pursue it. His research was funded and fueled by livestock deaths, not homicides. The flies in question were themselves suspected of the crimes. Though Hall was by no means the only dipterist on the case, it was in large part his legendary attention to detail that brought the culprits and their fascinating lifestyle into focus.

* * *

NATURE NEVER EQUIPPED a member of the Sarcophagid or Calliphorid families with the slicing, bladelike mandibles of the horse fly (family Tabanidae), the daggerlike mouthparts of a robber fly (family Asilidae), or even the piercing proboscis of a mosquito (family Culicidae). The husky bombardiers we know as blow and flesh flies spend their adult lives careening between flowers and overripe fruit, lapping up nectar and dripping juices as innocently as a butterfly. Occasionally one will commandeer a herd of aphids to steal a few drops of their sweet honeydew, and warm weather draws hordes to fresh excrement, where they dine on the sugary by-products of decay until disturbed into a buzzing, shimmering, blue-green swarm. Whatever their table of the moment, these noisy, iridescent flies feed with mouthparts as soft and spongy as that of a harmless house fly.

Yet combine all the bloodthirsty slashing and stabbing of every biting insect known in North America, and it would pale in comparison to the devastation blow and flesh flies were wreaking on U.S. cattle and sheep at the dawn of the twentieth century—most cruelly in the southern states. For the flies do not begin life as vegetarians. Quite the opposite, each mother supplies her young with a cradle of raw flesh.

Few if any American entomologists of that day appreciated the valuable service provided by the many blow and flesh fly species that infest only the dead and, in doing so, clean forest and field of its disease-carrying carrion. Their sole concern lay in eradicating the handful of calliphorid and sarcophagid flies that don't wait until their meat has lost its desire to swat. The most infamous of all was the ravenous blow fly maggot known as the screwworm, *Cochliomyia spp*. By the late 1920s, livestock losses to screwworm infestations were running as high as $10 million a year in Texas alone, despite rival sums spent on extermination.

Nor was myiasis, or fly infestation, strictly a veterinary problem. As far back as the late 1800s, Charles Valentine Riley, the founding curator of the National Museum's Department of Entomology, received regular correspondence from country doctors desperate for help, as in a letter from Dr. Fred Humbert, of Alton, Illinois:

> A farmer's wife, 35 years of age, was attacked on Monday, September 27, 1875, with a headache and flushed face. From this time, the pains in the region of the frontal cavity at the base of the nose and below the eye extending to the right ear increased. At times the pain was more severe than at others, but it never entirely left. This pain was described as preventing hearing and breathing and so excruciating that at intervals day and night her cries could be heard at a great distance from the house.
>
> Tuesday evening blood mucus began to run from the right nostril which was somewhat swollen, the swelling extending by Friday over the whole front side of her face. On this day, the fifth of the complaint, four large maggots dropped out of the right nostril. When I was first called to the patient, Monday, October 4, only the right lip and nostril were swollen, the acrid discharge having somewhat blistered the lip below. After each discharge, maggots dropped from the nostril, until the twelfth day, 140 or more maggots having escaped.

> On Monday, September 18, 1882, I saw a patient in the same neighborhood suffering from the same malady. At that time 280 maggots had been discharged . . .

By 1916, John Aldrich, the Smithsonian's first fly expert and Hall's predecessor at the National Museum, realized he was dealing with a confusion of Old World and New World flies, literally hundreds of physically similar species. Yet their outward similarities—so detailed as to defy classification—could mask differences in lifestyles and eating preferences large enough to confuse eradication efforts.

After publishing his monumental *Catalogue of the North American Diptera* in 1905, Aldrich focused on sorting out the flesh flies he knew so well from his farm-boy youth. Over the next decade, he distinguished scores of previously unrecognized species based on differences in the male fly's near-microscopic genitalia. Aldrich literally showed North American entomologists how to identify flies by looking to see how each was "hung." Executing the painstaking illustrations for Aldrich's revised *Sarcophaga and Allies of North America* was his protégé and soon-to-be-successor at the Smithsonian, David Hall.

Hall tackled the even larger challenge of mastering the continent's profusion of blow flies, the family Calliphoridae. In addition to his skill with anatomical illustration, he discovered a knack for raising blow flies by the thousands for developmental studies. Among the most important goals of such studies was the discovery of the vulnerable periods in an insect's life cycle when an application of pesticides or other control method might prove most effective.

Designing his own maggot hotels out of ice-cream tubs, Hall reared his blow fly larvae on ground chuck beef, recording the conditions needed for healthy development and timing their passage through each life stage to adulthood. He kept his adults content with a sweet mash of sugar and bananas, filling the back corridors of the National Museum of Natural History with their buzzing cages and the associated pungent-sweet smell. When the inevitable escapees made administrators twitchy, the curator continued his studies at home. As meticulous in her homemaking as her husband was in his own discipline, Pauline failed to appreciate the household's new additions. Fly cages under the bed were bad enough, but she drew the line when her icebox began filling with sherbet cartons filled with pupa undergoing "winter" diapause.

While Hall worked out the life cycles of many important blow flies, other North American dipterists began cataloging the group's habits in the field. Across the United States and southern Canada, they camped out in fields, barns, and woods, where they tallied the relative numbers of different species found on infested livestock and assorted carrion. They tossed out dead rabbits, dogs, and mice and recorded which flies arrived first and which waited till the carcass grew rank with decay. They noted which flies hung back in shaded places, and which sought out direct sun, which thrived in the swelter of summer, and which predominated in the cool of spring and fall. They recorded the maximum and minimum temperatures at which each species ceased activity altogether. With greater detail, a picture emerged of the conditions that made blow fly and flesh fly infestations possible.

Early on in the research, it became obvious that these bulldogs of the fly world couldn't pierce the tenderest of skin. Whether infesting cattle or carrion, the females instinctively positioned their eggs in and around ready-made openings such as the mucous membranes of the nose, mouth, and eye. Yet these attractions paled in comparison to the allure of an open wound, be it a gaping gash or the slightest blood-stippled scratch. Indeed, screwworm flies had been seen found exploiting so much as a flea bite.

On arrival, the egg-laden female may pause for a prenatal, protein snack before getting down to business. Turning in delicate pirouettes, she touches and tastes the wound through *chemoreceptors,* or sensory cells, on the soles of her six feet and the ovipositor, or egg-laying organ, on the tip of her abdomen. As the urge to extrude her eggs becomes overwhelming, she chooses her spot. Over the next six minutes, she can lay up to 400 eggs, spreading them like a beige paste of Parmesan cheese, first directly over the wound, then in thin shingles on the skin around it.

The eggs begin hatching ten to twenty-four hours later, depending on the outside temperature. Each larva emerges the size of a pen nib and the shape of a creamy white hot dog. Born ravenous, each larva immediately begins tearing at the exposed flesh with a pair of minute, black mouth hooks. As they grow, the maggots take on their familiar screwlike appearance, tapering from a flattened rump to a pointed mouth, with a series of threadlike ridges circling the body from one end to the other.

Together the rapidly growing maggots burrow and gouge deep pockets, slicing through connective tissue like butter with the aide of a unique cocktail of collagen-splitting enzymes. Bacteria already present on the newly hatched larvae spread to the flesh and aid in its breakdown. Packed together cheek-to-cheek, so to speak, the larvae bury their pointy heads in their food while wriggling their flattened rumps in the air above. A close look at a north-facing maggot's south end reveals two dark spots—breathing organs called "anal spiracles." Periodically, the metabolic heat of the maggots' feeding frenzy sends a roiling mass tumbling back to the surface to cool, before plunging back headfirst to feed.

Male flesh flies and blow flies perch nearby. Remaining on the outskirts of the bacchanal, they wait their turn to rise in the air with virgin females and new mothers eager to mate again. The frenzy continues until darkness, heavy rainfall, or dropping temperatures force the adults to take shelter in surrounding vegetation, where they rest till freed again by light and warmth.

As the adults swarm above them, the maggots grow quickly, laying down an opaque layer of body fat that gradually hides all internal organs save the crop, or feeding tube, that remains visible as a dark red line extending back from the mouth. In a span of four to seven days, they mature through three life stages, or instars, each larger than the last. At the end of the last and longest stage, the satiated larvae stop feeding. Each about a half-inch long, the mature, third instars pull themselves out of the roiling, crackling pack and drop to the ground. Instinctively, they seek darkness, swinging their light-sensitive heads left and right as they crawl.

Some hide beneath the fallen animal or under leaf litter and other debris. Most maggots bury themselves in the soil. There they contract their bodies into short, thick plugs and wait for their larval skins to harden around them. Inside their darkening shells, or puparium, they undergo their final metamorphosis into adults.

Depending on the season, temperature, and species, a new fly begins to stir inside the puparium anywhere from a week to six months later. Their emergence begins freakishly, with a fleshy sac called a *ptilinum* popping out of a slit above the adult's still sightless eyes. Rapidly filling with fluid, the sac presses against the top of the pupa case, popping it open along a fracture line. Like a pulsing water balloon, the ptilinum pulls the soft, pale protofly out of its unzipped sleeping bag and up through the soil to the surface. Emerging en masse, the new adults skitter across the ground like hyperactive spiders, waiting for their wings to unfold and their body cavities to fill with air. Launching themselves into the air some thirty minutes later, they begin the cycle anew.

Entomologists came to realize that live-feeding blow fly maggots such as the screwworm have some additional tricks. Their remarkable digestive juices included toxins that prevented livestock wounds from healing. So the first larvae to hatch on a tiny scratch could quickly enlarge the feeding area for siblings to come. The screwworm's damage also rang the dinner bell for its more numerous *necrophilous*, or "dead-loving," cousins. As bacterial infection takes hold of a wound, its rotting tissues begin to emit a bouquet of putrid odors, the unmistakable scent of death.

Scientists now know that nature honed the senses of these cadaver-feeding species to detect the molecules of decomposition in quantities as minute as a few parts per million. Blow flies and flesh flies "smell" acetic, butyric, and valeric acids, indole, acetone, phenol, and methyl disulfide through the many prominent body hairs that give them their distinctive, whiskery look. Chemical receptors on these special hairs, or *sensilla trichoidea,* lock onto the airborne molecules one by one, guiding the flies ever deeper into the vortex of chemical concentration, till they find themselves at ground zero, the source of the bewitching fragrance. This extreme sensitivity to the molecules of death makes perfect sense for an organism that must take advantage of an unpredictable and fleeting resource. When a corpse or carcass hits the ground, the flies must find it in the hours before decomposition begins turning flesh to mush. So it was that many of these beneficial "recyclers" ended up in bad company, elbowing for room in the festering wounds of America's sheep and cattle.

Ranchers and entomologists of the early twentieth century had no way of distinguishing the real livestock pests—that is, the blow and flesh flies that fed on healthy flesh—from harmless "hangers-on" who came only to trim away the dead and infected tissue. Indeed, the latter performed a genuine service first appreciated by Confederate surgeons, who successfully used bluebottle and greenbottle maggots to debride and disinfect the wounds of Civil War soldiers.

Such subtleties were lost on the desperate ranchers of the 1920s, who tried to stem their losses with folk remedies, plastering nicks and scrapes with soot and the sticky pine tar that they scraped from turpentine stills. Newly commissioned agricultural agents wisely advised ranchers to postpone bloody chores like dehorning, castrating, and tail-docking till winter, when cold weather would help keep flies in check—at least north of the frost line. But the battle was a losing one in the South, where fly infestations continued year-round.

Hope of salvation came in the early 1930s, when research confirmed what many had suspected. The screwworm fly, with its singular nose for the slightest scent of warm, healthy blood, lay at the heart of North America's staggering livestock losses. Stop this one species, and most likely, its scavenger cousins would lose interest in cows and sheep and go back to their carrion-breeding ways. Still, it seemed strange that although the screwworm could be found in virtually every corner of the world, only in North and Central America did it devastate livestock—most notably south of San Antonio, where mild winters failed to break the cycle of year-round infestation.

Armed with over a decade of intensive blow fly research, in 1932 the USDA launched a screwworm eradication program of a magnitude unmatched in agricultural entomology. Under the direction of agricultural agents, ranchers baited enormous fly traps, many the size of revival tents, with sacrificial cows, slaughtered by the hundreds and drug into place by teams of horses. In its first heady months, the campaign appeared to be a smashing success. Smug entomologists and ranchers posed for newspapermen in front of towering piles of dead flies some twenty feet high.

But as the country spiraled deeper into the Depression, government entomologists stood by dumbfounded as livestock losses continued to mount. It flew in the face of common sense and basic biology. The removal of hundreds of millions of screwworm flies brought not the slightest reduction in livestock myiasis. Ironically, the near collapse of the U.S. screwworm eradication program led to its ultimate success.

Furloughed during the barest bones years of the Depression, USDA entomologists Perry Cushing and William Patton left for England to pursue doctoral studies at the School of Tropical Medicine in Liverpool—taking with them several specimens of screwworm flies from the southern United States. Comparing their specimens to those in English collections, Cushing and Patton came to realize their colossal oversight. Minute differences in male genitalia revealed that North America had not one species of screwworm, but two. Not even Hall, with his legendary eye for detail, had seen the distinction.

The species that eventually proved the culprit behind 90 percent of myiasis cases turned out to be a relatively uncommon screwworm unique to the Americas, *Cochliomyia hominivorax*. Dwarfing it in numbers was an outwardly identical, carrion-eating screwworm, *C. macellaria,* which may have originated in the Old World. It was *C. macellaria* that had filled the USDA's carcass-baited traps to overflowing, whereas its livestock-killing cousin ignored the cold meat entirely. Entomologists named the newly distinguished *C. macellaria* the secondary screwworm, reflecting its secondary, or minor, role in wound infestation.

Another thirty years of research would reveal a chink in the lifestyle of the primary screwworm that ultimately proved its undoing. Female primary screwworms flies mate only once. In 1962, USDA entomologists devised a way to sterilize laboratory-raised males with sublethal doses of radiation. When released into infested areas by the millions each spring, the sterilized males ensure that the vast majority of females lay infertile eggs. Meanwhile, over a half century of exhaustive research on North America's flesh and blow flies supplied a host of well-described, postmortem clocks to the country's nascent field of forensic entomology.

In another twist of fate—at the same time that blow fly research reached full throttle—the National Museum of Natural History got a new neighbor. On November 24, 1932, the Federal Bureau of Investigation officially opened its Scientific Crime Detection Laboratory directly across the street in the Southern Railway Building. Crowded into a converted lounge, the makeshift laboratory consisted of a collection of gee-whiz gadgetry including an ultraviolet lamp for illuminating hidden fingerprints, an X-ray machine for peering inside suspicious packages, and a device dubbed the "helixometer," which supposedly allowed agents to examine the inside of gun barrels, but mostly served to impress visitors. Bugs were definitely not what J. Edgar Hoover had in mind when he christened his new public relations vehicle. Destined to be the finest crime lab in the world, the FBI sci-crime unit began as more pomp than nitty-gritty science.

Just as these dark-suited G-men started frequenting the Smithsonian's physical anthropologists, they likewise began appearing at Hall's door. As the National Museum's curator of diptera, Hall counted among his unavoidable duties the routine identification of fly specimens sent to him from across the country. Admittedly, most of this dreck work came by way of agricultural agents. But the vials of insects the FBI men pulled from their pockets differed only in the consistency with which they were traced to the two fly families that were Hall's specialty. The agents wanted to know if the maggots might somehow suggest when and where a body in question had become, well, fly bait.

They had certainly come to the right person. When it came to discerning one superficially identical species of blow or flesh fly from another, Hall had few equals. Not that the agents were in the habit of bringing identifiable adult flies. These guys packed Colt 38s, not butterfly nets. But every now and then, an agent desperate enough for a lead would stoop to pluck a few maggots off the body in question. So larvae, typically dead larvae, were what Hall got, often shriveled or otherwise poorly preserved. With a nod and a wave, he would tell the agents to come back the next day, asking only that they phone him with a local weather report for the days preceding the body's discovery.

To most of the world, a maggot is a maggot, but Hall didn't even need a magnifying glass to distinguish the characteristic cigar-shaped trunk, pointy head, and slightly upturned, concave rump that identified the larvae of the closely related flesh flies and blow flies. To tell the two groups apart, however, he needed a closer look at that distinctive rear end.

Had he been preparing a specimen for detailed, anatomical study, Hall would have painstakingly soaked it in a series of baths—alcohol, glacial acetic acid, oil of wintergreen—producing, over the course of two or more days, an almost transparent larva with its *sclerotinized,* or hardened, parts standing out in stark relief in their natural positions.

For routine identification work, Hall simply grasped the maggot with a pair of forceps and pushed its pointy end into a watch glass filled with fine sand. Slipping the tiny dish under the lens of his dissecting scope, he focused on the two dark dots on the maggot's rear end—its *anal spiracles,* or breathing tubes.

Early in his career, Hall found himself mildly disgusted at this peculiar arrangement of a maggot's breathing apparatus. Where else in nature could one find a creature sucking air through "nostrils" arranged so close to its anus? Over the years, however, he'd come to appreciate nature's exquisite "wisdom." After all, what other form of life fed with its head burrowed so deeply in rotting flesh or excrement?

Closer to Hall's painterly sense of aesthetics were the spiracles' delicate structure and arrangement. When viewed under a magnifying power of 25, the anal spiracles of a blow fly maggot resemble nothing so much as a pair of scalloped sand dollars, each embossed with a cluster of lacy, petallike slits fanned around an inner "button." By comparison, the spiracles of a flesh fly maggot appear somewhat less symmetrical, with broken rims surrounding a coarse set of slitted cracks. Making quick note of these details, Hall would have the larvae placed into its proper family before the departing government agent had tipped his fedora to the wide-eyed secretaries in the outer office.

Hall's initial inspection likewise revealed the specimen's life stage, or instar. Conveniently, the anal spiracles of a mature, or third instar, maggot contain three serrated slits; those of a second instar have two; the first instar—one.

Next came the real eyestrain. If Hall was lucky, he would have a few chunky third instars with which to work. Only at this stage of development could he hope to find the excruciatingly small details that distinguish one blow fly or flesh fly species from another. Did the outer rim, or *peritreme,* of each spiracle project down between the inner slits? If so, he might be dealing with the greenbottle *Phaenicia coeruleiviridis.* Did the inner button appear somewhat indistinct, not entirely enclosed by the peritreme? If yes, he might be looking at a secondary screwworm. Were the thornlike tubercles on the outer edge of the flattened rump distinctly longer than one-half the width of one anal spiracle? Perhaps the black blow fly, *Phormia regina.*

Turning the larvae around, Hall teased apart its so-called head with a dissecting needle, revealing a dark, anvil-shaped bit of cartilage, the maggot's internal mouthparts, or *cephalopharyngeal skeleton.* Hall would note whether the top of the structure's "anvil" extended over its base, and if so, how far and how high. He checked for a windowlike opening in the base of the anvil, or alternately a pigmented spot. Were the "teeth" attached to the mouthparts simple hooks or compound barbs? Were they short and clawlike or long like a saber?

Tallying the details, Hall narrowed the field of choices and, as often as not, nailed the species. The rest was simple arithmetic. Pulling out a sheaf of developmental timetables, Hall counted backward through the hours and days required for the egg of a given species to hatch and the resulting larvae to mature through each of three instars. Depending on the species, hatching could take anywhere from fifteen to twenty-six hours at laboratory temperatures of about 72 degrees Fahrenheit. The first and second instars would last between eleven and forty-eight hours; the final third instar, between thirty-six and sixty hours. If the FBI came up with a temperature for the crime scene, Hall could round his time estimate up or down, for hotter or cooler conditions.

Farther down the maggot time line, Hall recognized when a fully mature third-instar maggot had come to the end of its feeding cycle. As the satiated larva began to wander in search of a dark, pupal hideaway, the previously distinct red line of its flesh-filled gullet faded and disappeared. Even older was the maggot that had begun to contract its tapered body into the short, squat form of a prepupa, ready for its soft, creamy skin to stiffen and tan into a hard, dark pupal pellet.

Hall had a few other tricks up his sleeve as well. He understood that the moment a fly larva molted into a given life stage, its stiff, cartilaginous parts were set in stone, but the rest of its fleshy body continued to expand. So Hall could refine his age estimates by comparing the relative size of these hard structures—primarily the internal mouthparts and anal spiracles—with the larva's overall length.

Best of all were fortuitous moments when Hall looked through his microscope to see a sort of double image, with the shadowy outlines of one set of spiracles and mouth hooks taking form beneath a slightly smaller set. In those instances, Hall knew he had caught the larva in the act of molting from one instar to the next, a specific time in the life of a maggot.

When all else failed—as was the case with maggots so badly preserved as to be featureless—Hall simply measured them rump-to-mouth and compared their length to published information on the average, day-by-day growth rates of common fly larvae.

By the time the government agent sauntered back across Constitution Avenue the next day, the curator of diptera would have his report, seldom more than a paragraph, neatly typed. If everything had fallen neatly into place—well-preserved specimens, identifiable species, reliable weather information—Hall might narrow the time window when blow fly met body to a specific date. More often, he dared no more than a span of three to five days.

Always, Hall's report cautioned that his time estimate provided only a *minimum* postmortem interval. That is, he could say with certainty only that the corpse on which the maggots were found must have been dead and available for *at least* so many hours or days. Numerous factors, from cold weather to closed doors, could have delayed the blow flies' arrival.

If Hall felt confident in his species identification, his report might also offer some insight as to *where* the blow flies first sniffed out the corpse. This was the crux of what many agents wanted, as their cases often involved bodies found stuffed in crates, luggage, and car trunks—likely moved from the murder scene, perhaps across one or more state lines.

The presence of the greenbottle *Phaenicia sericata,* for example, might suggest to Hall that the body originated in an urban setting, but one that was open and sunlit, perhaps a rooftop or derelict lot. By contrast, the black blow fly, *Phormia regina,* predominated in rural areas. The greenbottle *Phaenicia eximia,* in turn, pointed decisively to the southern states, and the Holarctic blow fly, *Protophormia terraenovae,* generally kept to the North.

Thanking Hall for his time, the agents disappeared again as mysteriously as they'd come. So sporadic were their visits, never involving the same investigator twice, that Hall never knew how useful his determinations proved to be, nor did he ever ask for the grisly details of incoming requests. Still, he couldn't help but wonder.

From the almost daily barrage of gangland headlines, everyone knew the FBI was knee-deep in bodies. The media-savvy Hoover also kept the newspapers abreast of the scientific prowess with which his agency matched wits with underworld cunning. Technological wonders such as ballistics, blood typing, fiber analysis, and the "newly perfected" polygraph machine quickly captured the public's fancy. That "bugs on bodies" never made it through Hoover's public relations machine came as no surprise to Hall. Even he understood: Squirming maggots don't make good copy.

Nonetheless, Hall added "forensic applications" to his list of reasons for writing *The Blowflies of North America,* the book with which he planned to crown his career. Hall completed the historic monograph on *Calliphoridae* in 1948, after returning from service as an army medical entomologist during World War II. Soon after, he found himself effectively crippled as a taxonomist by cataracts. No longer able to discern the fine details on which he'd built a career,

Hall ceded his responsibilities at the National Museum in 1949 to Curtis Sabrosky, a jovial dipterist with no particular passion for blow flies, but an engaging way with the FBI agents and Washington, D.C., homicide detectives, whose visits were becoming routine. If they were going to waste his time, Sabrosky chided good-naturedly, they'd better learn how to make a decent collection.

For starters, Sabrosky wanted the investigators to bring him more generous samples. He told them to search "natural openings" such as eyes, nose, and mouth as well as wounds for even the smallest larvae. "Then bring me some of every size you see." A corpse supported a mixed community, he tried to explain. The greater the variety of "witnesses" at his disposal, the more information he might be able to retrieve. If that request didn't send a detective reeling, Sabrosky pressed further: Divide each sample into two parts, and stick half in alcohol—ideally a 70 to 80 percent solution of ethyl alcohol, but a shot of cheap vodka or whisky would do. Then keep the rest alive, each little brood in its own jar or bottle, and bring them to the museum for rearing into identifiable adults.

To say that Sabrosky's advice wasn't heeded would be an understatement. Macho types don't do well with maggots, he could have concluded, and left it at that. After all, his work with the FBI and local law enforcement only detracted from his own, undeniably valuable research on tachinid flies, a family with the beneficial habit of brooding their maggots inside caterpillars and other insect pests.

Still, Sabrosky didn't begrudge the diversion, and occasionally brought home sketchy accounts of the otherworldly violence that spilled into his laboratory every few months. He, perhaps even more than Hall, could see the vast potential of the insect evidence being squandered at death scenes across the country.

In 1961 Sabrosky decided to educate J. Edgar Hoover, no less, with a personal memorandum titled "Maggots in Corpses—A Measure of Time of Death." Sabrosky's seven-page report to the FBI chief laid out the usefulness of insect evidence, detailed proper collection techniques, and offered up a general "who's who" of bugs on bodies in various stages of decay. The Smithsonian dipterist never received so much as an acknowledgment. From the lack of mention in FBI archives, it would appear that forensic entomology's first procedural guide and collecting manual went no further than Hoover's "circular file."

Sabrosky's final entry into the annals of forensic entomology proved more satisfying, if little more noticed. In 1966, a homicide detective handed Sabrosky two envelopes stuffed with evidence. Exhibit A consisted of a single puparium, found inside the barrel of a gun. Exhibit B comprised many more of the same puparia—found beneath the floor mat of a burned-out car.

As Sabrosky would later learn, police had found both the gun in question and the burned-out car some fifty feet apart at the bottom of a cliff in rural, southwestern Virginia. Inside the vehicle were the charred remains of a young woman, who on autopsy was shown to have died of a bullet to the head.

By the time the insect evidence made its way to Sabrosky's lab, the police already had a confession from the woman's boyfriend. But prosecutors weren't satisfied with his story of a passionate argument gone terribly wrong—that is, second-degree murder, a crime for which he'd likely serve no more than five years. As part of their case for a first-degree, or premeditated, murder conviction, they hoped to punch holes in the boyfriend's anguished story, beginning with his claim that in his distraught state the night of the killing, he had stumbled away from the scene, oblivious to where he had rolled the car off the cliff. Quite the contrary, they suspected that the accused had come back later to try to conceal his crime by deliberately burning his girlfriend's body beyond recognition.

Ray Gagne, one of Sabrosky's colleagues at the Smithsonian, remembered the elder dipterist's unabashed delight at receiving a subpoena to actually testify at the trial. "He decided to take his wife and make a little vacation of it," Gagne recalled. And so on a pleasant morning in early summer, the couple drove five hours through the blossoming Virginia countryside to a small courthouse on the other side of the Blue Ridge Mountains. Mrs. Sabrosky sat in on the first half of the court case while her husband, barred until his testimony was called, sat outside in the courtroom hallway.

Calling their expert to the stand, the prosecutors invited Sabrosky to briefly educate the jury about the life stages of a maturing fly. He touched lightly on the blow fly's immediate attraction for fresh remains, the three life stages of the feeding maggot, and finally its migration off the body to find a dark place to pupate and metamorphize into an adult fly.

Did the puparium inside the gun match those found under the car floor mat in terms of age and species? a prosecuting attorney asked.

Yes, Sabrosky replied.

What could he, as a fly expert, conclude from this?

That the gun had been inside the car when the larvae finished their feeding and migrated off the body to seek a dark hideaway.

A dark hideaway such as a gun barrel?

Exactly.

Could a migrating maggot have found its way into the gun barrel if, as the defense suggests, this gun dropped out of the falling vehicle the night of the crime, landing some fifty feet from the car?

No, Sabrosky told the court. Any maggot leaving the car would have burrowed into the soil or leaf litter within a couple feet, a yard at most.

With this testimony, the prosecution convinced the jury that the suspect, far from being so discombobulated that he did not know where he killed his girlfriend, had in fact deliberately returned to the scene to disguise all signs of his involvement. They got their conviction for homicide in the first degree, and Sabrosky, to no particular acclaim, became the first North American entomologist to take the stand as material witness to murder.

6 A Model for Murder

Who saw him die?
"I," said the Fly,
"With my little eye,
I saw him die!"
—"WHO KILLED COCK ROBIN?"
(MOTHER GOOSE)

HAD THE BELEAGUERED Thaddeus Harris remained to haunt Harvard Yard a century after his death, he might have been shocked by the free love and protest rallies flowing over its grassy commons. But he would have been absolutely electrified by the entomology. Though social upheaval engulfed U.S. campuses in the 1960s, quieter forces were advancing insect studies beyond Harris's wildest dreams.

The sheer massiveness of the continent's insect fauna still promised discoveries in perpetuity. Nevertheless, U.S. and Canadian entomologists had sped far beyond the mere identification of new species to pioneer research in insect physiology, biochemistry, ecology, and behavior. North American scientists had likewise taken the lead in the forensic sciences. Between 1960 and 1974, they established the use of polarizing light microscopy for the identification of man-made fibers, gas chromatography for the analysis of arson chemicals, scanning electron microscopy with electron dispersive X rays for the detection of gunshot residues, radioimmunoassay techniques for screening blood and urine for narcotics, and gel-based isoenzyme tests for matching suspects to bloodstains and semen.

Yet for all practical purposes, the terms "forensic" and "entomology" remained as disconnected as ever. Yes, resourceful federal agents and homicide detectives still wandered into natural history museums and universities on occasion, looking for experts to identify something buglike at the scene of a crime, but no more so than they might consult a hardware clerk to identify a paint chip ("Definitely Sherwin-Williams Sunflower Cream").

Nor had any crime lab, homicide squad, or medical examiner added insects to the checklist of evidence routinely collected at murder scenes and criminal autopsies. With rare exception, flies and maggots remained revolting annoyances to be flicked away or doused with roach spray. This practice prevailed, despite the fact that not one of the high-tech gadgets piling up in crime labs and medical examiner offices could provide a reliable and consistent answer to the most basic of questions in any death investigation: When did the death take place?

The disconnect seems all the more baffling given pathology's helplessness in estimating time of death beyond the first twenty-four to forty-eight hours. Why hadn't entomology stepped in to fill the forensic gap? By 1960, nearly a dozen European and American entomologists had helped solve as many homicides with their identifications. Decades of research on blow flies and flesh flies had produced an ideal set of postmortem clocks. Yet no one seemed to see the larger picture, or wanted anything to do with it if they did.

Perhaps it was too revolting, at the most primal level. What police detective would want to make a habit of scooping up maggots? What entomologist would want to probe the infested orifices of murder victims? Besides, solving crimes with bugs in an age of spectral analysis and X-ray diffractometry must have seemed as anachronistic as treating blood infection with leeches.

But something began to shift over the next two decades. Perhaps the counterculture era had blurred the limits of what society deemed acceptable. Certainly, its back-to-nature ethic ushered in a new appreciation of the interconnectedness of all life—great and small, sublime and vile. In any case, between the mid-1960s and the early 1980s, three Americans would pick up the gauntlet thrown down by France's Mégnin nearly a century before and create a bona fide field of study. They would do so with shocking yet scientifically rigorous research that would make the worlds of both entomology and forensic science sit up and take notice.

* * *

JERRY PAYNE WAS a renegade even by the counterculture standards of the 1960s. A brilliant loner from the mountains of north Virginia, he came to South Carolina's Clemson College in 1961, wanting nothing so much as to find graduate work he could pursue in stubborn solitude. Payne's graduate adviser, the renowned entomologist Edwin King, steered his student away from the obvious choice—the ascetic refuge of taxonomy.

"No jobs, son."

King suggested instead the up-and-coming field of insect ecology. Indeed, he saw in the rough-hewn mountain boy a born naturalist. As Payne mulled his adviser's suggestion, he thought back to a project he'd completed for an undergraduate ecology course. Having read that carrion was a neglected "microsere"—that is, a small and fleeting ecological community—he'd slaughtered a couple of chickens and laid them beside a pair of dead frogs in a screened cage. The project's less-than-astonishing conclusion: Things mummify quicker without feathers.

Still, the dried-out remains had started Payne thinking about a larger enigma. Diseased and worn-out animals drop dead every day. Yet he seldom tripped over them when walking in the woods. Exactly who or what was speeding the cleanup? And what might alter the speed?

Payne suspected that insects played the major role and that their absence might cripple the entire process. He wondered if he could turn his hypothesis into a master's thesis. Reviewing the literature, he found precious little to quantify the impact and importance of nature's scavengers and decomposers. In fact, the small number of species identified as playing this role could not possibly carry out the Herculean task alone. With a private smile, Payne thought back to the wrinkled noses and disgusted groans his mummified chickens and frogs had provoked in his undergraduate classmates. In that moment, he realized he had found the kind of research he could pursue without anyone looking over his shoulder.

Fortunately, Payne brought much more to his work on insect decomposers than a reclusive attitude. Part and parcel of his almost hermetic personality was a staggering devotion to detail. Payne began by designing an elaborate version of Redi's seventeenth-century experiments in spontaneous generation. From mesh screen and wire, he built two sets of hinge-topped, open-bottomed cages—one insect-proof, one insect-open—to hold his test carcasses. Both designs would be sturdy enough to exclude larger scavengers such as possums, stray dogs, and children with sticks. Within its cage, each carcass would lay atop a square of nylon mesh so it could be lifted for weighing at regular intervals without disturbing the insects at work. With each weighing, Payne would also record temperature—in, on, under, and around each carcass—as well as describe the progress of decomposition in terms of odors, fluid production, bloating, and so forth.

For his study area, Payne selected a mixed hardwood-pine forest a quarter-mile hike from the college insectary, where he would mount and identify his collections. Importantly, the site stood within 300 yards of a weather station that would provide daily readings of outdoor temperature, humidity, and rainfall.

Payne's initial plans involved placing eight test cages at widely spaced intervals on the forest floor. But already his mind was racing with possible variations on the theme: cages hung from trees, buried in the ground, submerged in water, half-submerged, cages open only to small insects . . .

The final and most crucial decision for Payne was his choice of carrion model. He had tried a variety of roadkill, but concluded he would have difficulty getting specimens of consistent size and sufficient quantity. Besides, feathers and fur blocked his view of crawling and burrowing insects and mites. After months of deliberation, Payne asked his faculty advisers to approve the use of piglets, arguing that local farmers were willing to give him stillborns as well as newborns accidentally crushed by their mothers.

Payne chose not to elaborate on the other reason he considered pigs the ideal test subject: Their soft, nearly hairless skin very closely approximates that of a human. Payne clearly saw the implications. "It wasn't the kind of thing that anyone would discuss openly at the time," he later recalled. "But I was convinced of the practical applications. In fact, I was sure the FBI and the military had been secretly conducting research on human decomposition, if only I could get my hands on it."

Payne had already discovered the pioneering work of nineteenth-century forensic entomologists. When he failed to locate an English version of Mégnin's *La faune des cadavres,* he enrolled in a French class on the condition that he could make its translation his term project. He also struck up a correspondence with Belgian physician and self-taught entomologist Marcel Leclerq, who in the tradition of Bergeret had quietly kept the science alive in postwar Europe.

Payne also began a letter-writing campaign designed to shake loose covert research on humans in the United States. When his missives to J. Edgar Hoover failed to produce results, he enlisted the help of influential South Carolina senator Strom Thurmon. Still, nothing. Letters to military agencies proved no more fruitful. However, several military contacts expressed interest in Payne's research. One correspondent even urged Payne to dress some of his pigs in tiny uniforms to see how clothing might alter results, but the young woman Payne half-jokingly approached to do the sewing was not amused.

Over the summer of 1962, Payne stood watch over his piglets day and night as he cataloged and compared the decay of their small bodies in their insect-proof and insect-open cages. The differences were startling. The piglets kept free of insects remained largely intact, save for water loss, even at five weeks. Only after three months did fungi finally begin to disintegrate their mummified remains. By contrast, the piglets exposed to insects rendered down to bleached bone and bits of dried skin within seven days—a 90 percent reduction in their original weight.

In all, Payne collected 382 different species of invertebrates from his dead pigs, 301 of them insects. Going beyond mere proof of the insects' importance as recyclers, he described in unprecedented detail the interwoven timing of their arrivals and departures over six distinct stages of postmortem decay—fresh, bloated, active decay, advanced decay, dry decay, and remains. Payne did so because he understood, from his knowledge of insect physiology and from reading pathology textbooks, that the invisible gases formed at different stages of decomposition would attract different kinds of insects. Given an insect's keen sensitivity to the odors of its specific food source, Payne had great faith in its reliability as a marker for time since death.

In essence, Payne was replicating Mégnin's bugs-on-cadavers research for North American insects, moving it out of the realm of mere observation to scientifically controlled study. Mégnin, after all, never knew exactly how long the corpses he saw had been dead, or how freely insects had access to them. By contrast, Payne's scientific methods would enable him to document time of death, control insect access, and note the start and finish of each stage of decomposition.

Payne had planned to mark the beginning of the "fresh" stage with the thawing of his pigs, for he had asked his farmers to freeze them immediately after their stillbirth or death. But on summer mornings, flesh flies began exploring and laying eggs around the nostrils, eyes, mouth, umbilical cord, and anus of each pig even before the flesh had a chance to shed its last ice crystals. Payne could only guess at what had attracted their attention so early in the decomposition process. At this stage, he could see no sign of decay or smell anything but the mild aroma of straw and pig mash that still clung to each small carcass. He identified the majority of the first arrivers as various species of flesh flies, *Sarcophaga spp.* By afternoon, they had been joined by a half-dozen blow flies including secondary screwworms, *Cochliomyia macellaria;* greenbottles, *Phaenicia coeruleiviridis;* and black blow flies, *Phormia regina.* Yellow jackets, *Vespula maculifrons,* and bald-faced hornets, *Vespula maculata,* arrived concurrently to feed on both the piglets' tender skin and the adult flies. Soon after, ants, *Formica spp.,* discovered the fresh remains and swarmed over the still-pink flesh to pull off the first fly eggs. As decay fluids began to drip from the piglets late in the day, house flies, *Musca domestica,* arrived to feed on the fluids.

The pigs entered the "bloat" stage the second day, as the bacteria multiplying within their guts began producing an abundance of methane, sulfur dioxide, and other gaseous by-products. Now Payne could see and smell the signs of decomposition—beginning with a slight inflation of the piglets' bellies early in the morning. The associated scent—still barely perceptible to Payne's nose, drew thick swarms of flesh and blow flies, as well as cheese skippers, *Piophila casei,* and fruit and vinegar flies, *Drosophila spp.* By afternoon, the buildup of internal gases began purging frothy fluids from snout and anus. The bloat stage continued over an average of two days, during which time the skunklike smell of putrefaction, or bacterial decay, permeated Payne's study area. By the third morning, the adult blow flies and flesh flies had lost interest in the remains, even as their recently hatched larvae spread out to feed on the face, belly, and genital areas.

Payne marked the beginning of his third stage, "active decay," on day 4, as the ravenous fly larvae broke through to the abdominal cavity to release the internal gases and rapidly deflate the small carcasses. As the decay liquids soaked into the underlying soil, they drew the attention of slender rove beetles, *Platydracus spp.*, and buttonlike clown beetles, *Hister spp.*, which quickly buried themselves beneath the bloated carcasses. From time to time, the beetles would scuttle into view long enough to grab a few tender maggots, before disappearing again beneath the remains. Simultaneously, the near-overwhelming stench of the released decay gases drew clouds of house flies to the scene. Active decay continued through the fifth day, when the maggots formed packs that actively swarmed through the chest and abdominal cavities. Over the course of the day, the piglets appeared to liquefy before Payne's eyes, as fluids and semisolid tissues flowed into the dirt. The sweet smell of fermentation drew a variety of butterflies, bees, and moths.

Payne marked the beginning of "advanced decay" on day 6, with the disappearance of most soft tissue. The odors of decay began to fade except for a lingering ammonia smell. At the same time, the fattened, third-instar maggots began abandoning the hollowed-out carcasses en masse. At times, their jostling grew so intense that it moved the remains one to two feet. Still more beetles arrived to feast on the migrating maggots. Advanced decay continued for a second day, as the carcass took on a dry and tattered appearance and the predaceous beetles departed with the last of the maggots.

Between days 7 and 8, Payne noted that each carcass entered the fifth stage of "dry decay," with a noticeable odor shift to that of wet fur and old leather. A few of the piglets still retained their general form and shape. Others consisted of no more than a scattered and entangled mass of bones, cartilage, and skin. He saw latrine fly maggots, *Chrysomya megacephala,* hatching, but lacking food, they failed to complete their development. Over the next few days, as the last stragglers from earlier stages disappeared, new groups of insects arrived to gnaw on the dried remains—skin beetles, *Dermestes spp.,* hide beetles, *Trox spp.,* checkered beetles, *Necrobia spp.,* and a variety of mites and roaches. Spading the ground under and around the remains, Payne found an abundance of tiny soil beetles and springtails feeding on the greasy seepage. Other dry-stage insects included a variety of gnats that Payne suspected were feeding on the fungi that now coated the remains.

Decomposition had downshifted into a period of slow molder that continued for a summer month before gradually blending into the sixth and final stage of "remains"—a near-odorless mat of hair, bits of dry skin, bleached bones, and teeth. Centipedes and millipedes took up residence beneath the bones. Ants periodically returned to carry away dry scraps. Occasional summer showers temporarily reversed the succession by making the remains attractive again to a variety of small flies and beetles.

Overall, Payne noted several important outside influences on insect activity. The flies, in particular, appeared keenly sensitive to changes in light. Dusk brought their activity to an utter standstill. Even a cloudy day triggered a noticeable sluggishness. Increasing daytime temperatures had the opposite effect, whipping insect activity to a frenzy on the hottest afternoons. The carcasses themselves produced heat, particularly during active and advanced decay, when they measured up to 16 degrees Celsius (29 degrees Fahrenheit), warmer than the surrounding air. The heat—presumably generated by bacterial and maggot activity—seemed to speed both larval growth and species succession. Moisture likewise affected activity. Excess fluids easily overwhelmed the smallest blow fly maggots, while wasps, butterflies, and certain beetles actually sought out the soupiest corners.

After earning his master's degree in 1963, Payne continued his bugs-on-piglet studies with a doctoral project that indulged his dream of stringing up, burying, and submerging carcasses to see how the changing circumstances affected their insect-assisted decay. Payne consciously designed the scenarios to replicate typical homicide and suicide scenarios. Once again, he judged that it would prove too scandalous to directly state the underlying purpose of his studies in "insect ecology." Even without mention of murder, the work proved titillating. A flurry of media interest in Payne's "insect morticians" followed the publications of his experimental results in the respected scientific journal *Ecology* in 1966. "If it weren't for insects," *Time* magazine reported, "Payne says we'd be up to our necks in dead bodies."

Payne's postgraduate job offers included a commission as second lieutenant with the Armed Services Institute of Pathology. Although he was tempted by the tacit invitation to pursue his postmortem research in battlefield situations, Payne concluded that his personality was no fit for the military. He accepted, instead, a position studying crop pests with the USDA, and his decomposition studies faded from view. They would remain forgotten in the dusty annals of obscure research for over a decade, until rediscovered, almost simultaneously, by two scientists in seemingly unrelated fields . . . two scientists who would make crystal clear the intent of their work, and in doing so, change the nature of homicide investigation forever.

* * *

EASTERN LOUISIANA ROLLS back from the giant lagoon of Lake Pontchartrain like a soggy green carpet under gray-milk skies. On a planet where insect development hinges very much on temperature and humidity, this patch of earth has plenty of both. In the words of Louisiana entomologist Abe Oliver, it is "one of the great insect incubators of the world."

In such a place, an entomologist can narrow his or her focus beyond just a group of insects to a particular developmental stage. For Oliver, an expert on the pests of ornamental shrubs, that was baby bugs—larvae. Unfortunately, the duties of teaching immature insect biology at Louisiana State University in the 1970s went hand in hand with the occasional visit from a parish detective toting a jar of unidentified maggots. As infrequent as they were—seldom more than once a year—they were visits Oliver detested.

"Anytime you put your eyes over a microscope and get that close, it's not going to be pleasant," he recalled. "The terrible odor. The visible bits of human flesh. . . . I considered the work a public service. But it was never kind."

So it was with keen personal interest that Oliver noted the newest addition to the department's faculty in September 1975. A former college football player, Lamar Meek fairly bounded through the halls of the university's agricultural college and filled its entomology lab with roars of laughter. Meek's ruddy complexion and tousled shock of dark hair only added to the advertisement for untapped energy. Having trained as a graduate student in mosquito ecology and control at Texas A&M, Meek had been hired ostensibly to pursue research on another family of flies—the tabanids, or horse and deer flies. But the department's senior faculty quickly recognized the new professor's greater potential—the only question was how best to channel it.

It wasn't merely one or two maggot identifications a year that Oliver hoped to pawn off on Meek. The elder professor had good reason to believe that the department's forensic caseload was about to increase dramatically. A few months earlier, Oliver had been pulled into a particularly gruesome and sensational double murder in nearby East Baton Rouge. Police had brought Oliver the maggots collected from two bodies found in an apartment house dumpster. Oliver's identification and aging of the larvae had cracked the high-profile case by busting the alibi of the prime suspect—the son of the local police chief. Word was out that the defense attorney claimed he could have countered every piece of evidence against his client except "those damn bugs." So impressed was the staff of the state crime lab that they resolved to make insect collection standard procedure in all future homicide investigations. Horrified, Oliver resolved to have no part of the flood of casework to come.

Department chair Dayton Steelman commiserated with Oliver, but saw the forensic assignments as more than nauseating dreck work to be pawned off on junior faculty members. Having himself provided a couple of time-of-death estimates based on maggots, he recognized the amount of guesswork going into these so-called scientific determinations—determinations that could put an innocent man on death row or free a cold-blooded killer. Steelman saw the pressing need for legitimate research. As the department's rainmaker, he also saw a potential spigot for grant money.

"It seems to me that crime labs all over the country would want someone who could teach police how to collect insect evidence," he told Meek a few days later. "Someone who can work with them and set up studies in the field." As Steelman had hoped, his eager young buck actually warmed to the idea.

Like any good scientist, Meek began his research with a hunt through the published literature, not that the term "forensic entomology" would appear in any scientific index of the day. What he found was Payne's ecological studies of insects on pig carrion. He tracked down the reclusive entomologist, then working at an agricultural research station in Tifton, Georgia. Unabashedly delighted to have his work finally recognized and appreciated for what it was, Payne showered Meek with information and encouragement. To replicate the decomposition and insect colonization of an actual murder victim, they agreed, Meek would need to size up the "Payne Pig Model" considerably. A thirty- to fifty-pound pig would be ideal, very closely approximating the torso of an adult human.

Meanwhile, Meek's grant proposals failed to turn into the cash cow the department had hoped. Yes, the Louisiana Department of Justice was keenly interested in Meek's research, and would pay him to train their homicide investigators. But research funds? The best the department could offer was the part-time use of Charlie Andrews, the crime technician who had brought Oliver the earlier maggots and who had been championing the use of insects on a statewide level ever since. The university, in turn, produced a part-time research assistant in the form of a grad student. But before Meek could organize his methods, a real-life experiment presented itself.

* * *

SOMEDAY THE WORLD would know her—or at least Hollywood's version of her—as Sean Penn's victim in the movie *Dead Man Walking.* But on an achingly beautiful spring morning in 1980, eighteen-year-old Faith Hathaway of Mandeville, Louisiana, was just a missing person. Admittedly, few people beyond Faith's parents held out hope of finding her alive.

Elizabeth and Vernon Hathaway reported their daughter missing on the afternoon of May 28. Faith, newly graduated from Mandeville High School, had disappeared literally on the eve of her new life, having gone to celebrate with friends before leaving "to see the world" with the army. On May 31, a bloodied and beaten teenager stumbled into the sheriff's office in nearby Madisonville, screaming that two men had abducted her and her boyfriend at gunpoint and driven them hundreds of miles over two days, repeatedly raping her in the backseat. In the end, they had stabbed and shot her boyfriend and left him for dead tied to a tree before dumping her on the side of the road. When the hysterical girl turned out *not* to be Hathaway, the police braced for what looked like a killing spree.

On Sunday, June 1, a family picnicking near Fricke's Cave, a remote swimming hole at the bottom of a steep, sandy gorge, found a purse and a torn blue blouse. They were Hathaway's. Search parties combed the surrounding area for two days, but failed to find anything more.

The morning of June 4, deputy investigator Michael Varnado woke determined to climb back into that gorge and not come out till he found Hathaway, or whatever was left of her. Varnado knew the spot well, having picnicked and swum there every summer as a boy. The swimming hole has pretty pink sand like you'll find nowhere else in Louisiana.

At the end of the rough dirt road, Varnado left his partner in the squad car, scrambled down the ravine, and pressed past the snarl of underbrush that separated "accessible" parts of the gorge from deep swamp. The idyllic spring morning had already given way to oppressive heat when he picked up the smell. Just four years on the force and Varnado already knew the odor instantly. What he *saw*, however, made no sense. A chalk outline on black mud . . . a stark, white tracing in the shape of a human body.

Then, it registered. The white line was moving. No . . . *churning*.

Dropping to his knees in shock, Varnado realized he was staring at the figure of Faith Hathaway, her decay-bronzed flesh almost invisible inside a bright circle of departing maggots. Slowly the details took shape: arms thrown back above her head; legs spread-eagled; more maggots churning across a nearly severed neck, over the blood-blackened chest, between the legs.

Pressing back to the trail, Varnado screamed up the gorge for his partner to radio the crime lab. Ninety minutes later, crime tech Charlie Andrews arrived on the scene, his eyes widening at the sight. "Mike," he told the young officer, "let me tell you what these maggots are going to do for us. . . ."

Over the next two hours, Andrews methodically processed the scene, starting with the insects. He collected short, squat prepupae from around the body and smaller, actively feeding larvae from the flesh itself. He plunged half of each batch in vials filled with a killing solution supplied by Meek. The others went into paper cartons for live rearing. Andrews labeled everything meticulously with date, hour, and collection site ("neck," "chest," "hands," and so on).

Meanwhile, police had arrested Robert Lee Willie and Joseph Vaccaro for the rape and kidnapping of the second girl. Miraculously, her boyfriend had survived his brutal beating, though it had left him partially paralyzed. When questioned, Willie and Vaccaro confessed to the couple's kidnapping, assault, and rape, but denied having anything to do with Hathaway.

Doubtful, Detective Varnado asked the kidnapped girl to drive with him down the long, rough road to Fricke's Cave. Gulping back her terror, she confirmed that it was the first stop in their rampage of terror. Her statement placed Willie and Vaccaro in the ravine on the night of May 29. If the police could show Hathaway had been killed there that night or the night before, they knew they had a case.

Andrews's exhaustive maggot collection was just what Meek needed. His calculations indicated that the first blow flies had laid their eggs on Hathaway's body seven days before its discovery—sometime between her disappearance on the night of May 28 and nightfall on May 29.

Confronted with the "coincidence," Willie and Vaccaro confessed. Shortly after midnight on May 28, they'd met Hathaway outside the bar where she'd been celebrating and offered her a ride home. Both recounted similar stories of blindfolding the girl, leading her into the gully, and raping her. But each blamed the other for the actual stabbing.

"Faith didn't know the animals she was dealing with," her father would lament. "You know young people. They think everybody's their friend."

On October 22, Meek made his debut as an expert witness at the state supreme courthouse, Washington Parish. Minutes after completing his testimony in *Louisiana v. Willie,* he sprinted upstairs to take the stand in *Louisiana v. Vaccaro*. Vaccaro drew a life sentence. Willie received the death penalty and posthumous fame, based on his death-row friendship with the activist Catholic nun Helen Prejean.

Shedding his three-piece suit for coveralls after the trial, Meek spent the following days hammering together six-foot-by-two-foot field cages for his first round of pig studies. Although he was pleased with his courtroom presentation, Meek fully realized that his testimony was vulnerable to attack by anyone who understood the subtleties of insect development. Like others before him, Meek had based his calculations on studies of blow flies reared in unnaturally ideal laboratory conditions. By contrast, the larvae developing on Hathaway's body would have had to contend with rain, humidity, heat, overcrowding, and a host of insect raiders. Only with field tests could Meek determine how such challenges could affect insect development in a variety of habitats and seasons.

Simply securing the necessary permits for his experiments had already consumed the good part of a year. Andrews's police firearm training proved crucial in finally winning the approval of the university's Animal Use and Care Committee, charged with ensuring that experimental animals suffered no undue pain. Andrews vouched that he could dispatch each pig instantly, with a bullet behind the ear. The ag college's Swine Center agreed to supply the pigs. Meek, Andrews, and grad student Mike Andis would then truck the freshly killed carcasses directly to their field sites a few miles away.

It was also Andrews who narrowed the choice of sites to best replicate homicides. Although murder motives may vary, body dumpings had a monotonously predictable pattern, and Andrews knew all the classic spots: dumped on the side of a rural highway or dirt road, buried in a shallow grave in the woods. For their field studies, he chose a deeply wooded site in a commercial forest and another alongside an open field near a university firing range.

So on a mild winter morning in January 1981, Meek and his small crew tossed out their first pigs, covering each with a scavenger-proof cage. They spent the day driving huge circles through their "carcass fields" to see how fast the flies pulled in. Within six hours, every pig lay beneath a paste of eggs save one. The odd exception proved instructive. Inadvertently, the men had laid an old sow atop a small fire ant mound. The aggressive ants had immediately swarmed over the body, pulling off fly eggs as fast as they were laid. The flies eventually established a toehold on the third day, starting their postmortem clock ticking a full seventy-two hours late. A fluke, perhaps, but a possibility Meek would have to keep in mind anywhere in the South.

Overall, the succession of insects roughly paralleled that documented by Payne, with expected differences in local species. Meek's collections produced a reference set with representative insects for each day and stage of decay. Beyond the order of arrival, Meek needed to nail down the duration of each "wave." Most important of all, for homicide investigations, would be the first surge of flies. If Meek had any hope of narrowing the time of death to a specific day or portion thereof, he needed to document how the flies' arrival and the maggots' development varied under different conditions of light, heat, humidity, and rain.

To this end, he continued to toss out pigs every few months to document how fly species and developmental times changed with the seasons. As expected, certain species such as the black blow fly predominated in the cooler months of fall and winter. Others, like the secondary screwworm, took over in the swelter of late spring. Growth rates rose with temperature, as Meek had expected. But even Meek found himself stupefied by the frenzy of summer. Over the course of four days in July, the maggots reduced a full-grown, thirty-pound hog to a five-pound mound of skin and bones. Virtually nothing remained for later waves of insects.

As Meek's studies progressed through the fall of 1981, Steelman became increasingly impressed with the new field of research materializing within his department. Taking Meek aside one afternoon, he asked his junior faculty member to prepare a talk and slide show for the upcoming meeting of the Entomological Society of America. Steelman had been asked to organize the medical entomology portion of the annual program, and he wanted Meek's work as his headline act. Clearly, it was not the kind of insects-as-disease-carriers research people had come to expect from such sessions. But Steelman had a deep-down feeling: Either this research would offend the hell out of everyone or it would launch a new entomological specialty.

What neither Meek nor Steelman could have imagined at that moment was that two anthropologists at the University of Tennessee were venturing into the same territory with an even more shocking model for murder.

* * *

BILL BASS COULD hardly believe his own audacity. It was not enough that he had already requested—and been granted—one holding facility for unburied corpses. Now he was back, eight years later, asking for another, and this one in town. Perhaps something on campus? All Bass's hopes for genuine research on determining time since death hinged on the chancellor's answer.

The idea of pursuing active research on his body collection had crept up on Bass during his periodic field trips to the anthropology department's sow barn near the headwaters of the Tennessee River. Bass had found it a peaceful place to spend a morning or afternoon, though the open shed that housed his unclaimed cadavers had become overcrowded. Still, he and his students often loitered after installing a new occupant. Some shared their professor's appreciation for the haunting beauty in the sag of cheek and jaw, the tawny darkening of mummified skin, or the soft pillow of gray hair that sloughs from an elderly head in the weeks after death. There were times when someone would sheepishly admit to having looked into a clouded set of eyes hoping for some insight into "the meaning of it all." But if some spark of profundity once inhabited these corpses, it had long departed before their waxy flesh ended up here. Just as Bass and his students saw nothing gruesome in their cadavers, neither did they see anything spiritually profound.

Yet on another level, every corpse seemed to brim with urgent secrets. There was something about the way steam rose from the bodies on a cold winter day, as the maggots teemed within their cavities, something about the lightning speed with which flesh melted from bone in the summer heat. At times, Bass felt *so close* to wringing order from the process of human decay. Then something would mystify him. Why did the flies swarm to one corpse and completely ignore another? Why did bacterial putrefaction consume one body in a matter of days and take weeks to render down another? More than once, he and his students had watched a 200-pound corpse virtually liquefy before their eyes, while a body half the size, just a couple feet away, took twice as long to decompose. Yet for all the maddening variation, Bass understood that he had a matchless opportunity to find order in the chaos, to find the variables that determine decay rates, and so solve the age-old enigma of how to calculate time since death.

He had at his fingertips a steady stream of bodies and a small army of graduate students to study them. What he did not have was a location conducive to daily observation. It took up to three hours to make the round trip along rugged backroads that led to the remote university farm, and that made active research impractical for professors and graduate students with demanding class schedules back on campus.

Amazingly, Chancellor Reese calmly listened to Bass's audacious request and promised to give it serious consideration. It did not hurt that, in less than a decade, Bass had built a world-class anthropology department from the ground up, generating more positive publicity than anyone could have imagined. Indeed, only the "Fighting Vols" football squad put the University of Tennessee in the national papers on a steadier basis.

In early spring of 1981, the chancellor gave Bass the use of three wooded acres behind the university's medical center, directly across the meandering Tennessee River from downtown Knoxville. The site encompassed a former school dump closed by the EPA's ban on open burning. It included a well-trampled field littered with rusting cars and surrounded by a wooded hillside tangled with underbrush and laced with deer trails. In short, it was the ideal habitat for replicating murder in all its most popular outdoor settings. Bass dubbed his al fresco mortuary the Anthropological Research Facility, ARF for short. (A criminal lawyer who later got wind of its purpose prefaced the name with the professor's to produce a more descriptive acronym.)

The plans Bass had for the place knew no bounds. He and his students would chart the effects on decay rates of temperature, humidity, rainfall, and soil pH. They could leave some bodies exposed, others buried at various depths, and still others locked in car trunks, stuffed in garbage bags, wrapped in blankets, naked, or heavily clothed. Bass had seen it all. Now he could turn the tables on the depravity and systemize it. Bass also wanted to document the decay variables that stemmed from variation in the human condition—height, weight, age, sex, cause of death, and manner of wounds.

First, Bass understood that he would need to begin simply and build a methodology that could stand up to scrutiny by the most vociferous of critics. There was no precedent for what they were about to do, in either forensics or anthropology. Their results would likewise have to stand up to legal challenges before they could ever be heard in a court of law. The question became *what* should they tackle first.

Intense and ambitious, graduate student William Rodriguez felt sure he had the perfect study. Rodriguez had come to anthropology by way of a master's degree in zoology. When he looked at a week-old corpse, that's exactly what he saw—a teaming zoo of creatures drawn to a highly nutritious, if fleeting, resource. Rodriguez wanted to do more than record what bone fell off when: He felt sure that insects played a decisive role, if not *the* decisive role, in driving decay. For his doctoral research, he wanted to study their potential as markers of postmortem interval. Given the overwhelming presence of maggots in his own forensic casework, Bass readily agreed.

The two anthropologists had no inkling of the parallel research unfolding at the same time in the humid forests and fields of Louisiana. Yet like Meek, they followed the paper trail to Payne's published studies on insect-assisted decomposition. Payne's findings that a carcass free of insects tended to slowly mummify confirmed Rodriguez's conviction of the insects' importance.

Together, Bass and Rodriguez ironed out the remaining details. To begin as close to ground zero as possible, each body would need to be within a few days of death, with any interim storage in a morgue cooler to keep decomposition on hold. They would place each subject faceup, so Rodriguez could easily record early insect activity around the eyes, nose, and mouth. They debated briefly over whether to leave clothing in place. However, since clothes would obscure their view, they would come off, and the legs would be placed slightly apart. Like the face and open wounds, the urogenital area was among the first areas Bass saw colonized by maggots in his homicide cases. Although he was no entomologist, Bass understood that the insects needed some kind of natural opening to take hold.

Finally, they would exclude larger, noninsect scavengers—raccoons, opossums, crows, and the like—at least at this stage of the research. Rodriguez hammered together several chicken-wire coffins, each with a lid hinged for easy access. Meanwhile, Bass had a concrete slab poured in an area of open woods where the subjects would lie in dappled light rather than direct sun. Around the sixteen-foot-square slab, they installed a chain-link fence with a chain-link roof to form a gated cage. They would set the wire coffins inside, elevated on corner posts just enough so Rodriguez could see beneath a body without moving it.

The year's first volunteer arrived in mid-May 1981, just as the tulip poplars were dropping the last of their powdery green blooms. Subject "1-81"—a seventy-three-year-old medical center patient who had donated his body to science—was drawing a cloud of flies within two hours of his arrival. Soon after, Rodriguez noted the first females settling down to paste their eggs in the moist corners of the cadaver's mouth, eyes, and nose. Rodriguez returned each day at noon with notebook, camera, butterfly net, and lunch. For two to three hours, he recorded and photographed 1-81's physical changes and collected insects for identification. Maggot activity had reached a peak when "2-81," another seventy-three-year-old male, arrived on June 5.

The work proceeded smoothly until summer introduced a variable no one had considered: idle schoolchildren. Unknown to the anthropologists, young hikers had visited their forest late one afternoon, having only to climb the low fence that surrounded the former school dump. What they told their parents is anyone's guess. The next day, a newspaper reporter called Bass's office, taking him by surprise. As was his way, Bass invited the journalist to come see what they were doing and learn why.

"I've already been," the reporter informed him. The kids had led him to the moldering bodies. The pictures, at least those fit to print, would be on the front page of the morning paper, alongside a file photo of Bass. Bass called the chancellor to forewarn him, and everyone in the department braced for the public reaction. Not surprisingly, outraged protests poured in to the school administration as well as the governor's office and the editorial pages of the *Knoxville News Sentinel.* Within the week, Bass was being reviled by talk show hosts on national radio for "this ghastly affront to human dignity." Medical ethicists labeled the work "Frankensteinish."

Bass and his graduate students feared their gig was up. To their amazement, the university issued a statement defending the research as providing valuable, potentially life-saving information. Tennessee's police chiefs, district attorneys, and medical examiners lined up behind Bass as well, describing to both the news media and the state legislature how they relied on the anthropologist's unique expertise to crack many difficult cases. On October 12, Bass installed a third male septuagenarian for Rodriguez's fall watch.

Meanwhile, Rodriguez found himself dealing with curiosity seekers. Weekends brought whole families hiking through the woods to gawk. Late nights drew frat boys, some soused, some trying to impress the giggling cheerleaders on their arms. The trespassing grew steadily worse with the approach of Halloween, till Rodriguez, fearing for his doctoral project, took to camping out alongside his subjects to chase off would-be vandals.

Making matters worse, the nearby university medical center had begun enlarging its employee parking lot in the direction of the research, leveling the forest cover on two sides. Protests flared anew when one of the tractor-grader drivers claimed he'd been "defiled" by flies he saw rising from one of the bodies. "They were all over me!" he told local TV reporters. Now Rodriguez had to fight his way past picketers and TV crews on his way to work. "THIS MAKES ME S.I.C.K.!!!" read the signs raised to the cameras. The Society for Issues of Concern to Knoxville had found an issue that would burn their acronym into the minds of Tennesseans for years to come.

Wisely, Chancellor Reese found the funds to build a chain-link fence topped with barbed wire around the three-acre research facility, the fence posts embedded in concrete. Soon after, the medical center sprang for a stockade "modesty fence" (eight feet high) inside the chain-link fence. The local health department nailed a red-and-white "Biohazard" sign on the double-padlocked gate. Peace had been largely restored by the time Rodriguez got his final subject, a thirty-five-year-old female traffic fatality, on November 11. By January, he had something to report.

Plainly, the human decomposition Rodriguez observed extended over a longer span of time than had the rapid decay of Payne's tiny piglets, though at the height of summer, the body of a full-grown man rendered down to dry bone in under a month. Yet the actual stages of decomposition remained similar, although Rodriguez truncated Payne's six stages into four—fresh, bloat, active decay, and dry remains.

More important, perhaps, Rodriguez collected the same insect families and many of the same species as Payne, although they differed somewhat in the timing and order of their arrival. In Payne's South Carolina studies, for example, flesh flies consistently arrived first, often within minutes of the entomologist slipping his frozen piglets from his cooler chest. Blow flies followed second, with muscid, or house flies, a distant third. By contrast, blow flies followed by muscids dominated the Tennessee cadavers throughout the fresh stage, with flesh flies not arriving until the putrid gases of microbial decay marked the bodies' passage into "bloat." Such variation might have to do with the difference between pig and human flesh, but Rodriguez doubted it. More likely, the differences lay in the idiosyncrasies of local insect populations, something that would demand caution whenever applying the results of a study in one part of the country to corpses in another.

Otherwise, the larger insect families that Rodriguez collected followed the same general pattern of succession that Payne had seen—flies in the fresh and bloat stage, followed by maggot predators such as clown and rove beetles during bloat and active decay, followed by a cleanup squad of dermestid and scarab beetles that fed on the dry remains.

Like Meek, Rodriguez was also struck by the powerful role temperature played in both decomposition rates and the arrival times and abundance of insects. In particular, he observed that the chill of winter performed much the same function as the finely screened cages that Payne had used to exclude insects from his "control" pigs. Without insects to speed their decomposition, Rodriguez's fall and winter cadavers slowly mummified.

Bass advised his student to use the winter holiday break to summarize his findings and put together a presentation for the February meeting of the American Academy of Forensic Science in Orlando, Florida. Bass also pulled the necessary strings to make sure his young apprentice would be assigned to the largest auditorium at the conference. "Don't worry," he assured Rodriguez. "Time of death always draws a crowd." It would prove an understatement.

Rumors of Bass's unconventional research facility preceded the UT anthropology department's arrival at the annual forensics conference. The morning of his presentation, Rodriguez had to press through a crush of police officers, crime technicians, medical examiners, and forensic anthropologists to reach the double doors of his lecture hall. Breakfast flipped in his stomach when he noticed that Bass was nowhere to be seen. Bass was there, of course, keeping a low profile in the back as he always did when one of his students made his or her debut.

At the lectern, Rodriguez waited in vain for the animated conversation to die down before launching into his prepared remarks. "The decomposing corpse, which has been discovered lying in a wooded area, secluded lot, or open field, presents the forensic scientist with many questions to be answered. One of the most important is the time interval since death." Rodriguez motioned for lowered lights and the first slide. "The cadavers of three adult white males and one adult white female were used in this study."

Sudden silence. Projected onto the screen was a wide shot of the chain-link cage, inside of which could be seen a chicken-wire coffin holding subject "1-81" in full bloat. Rodriguez heard a ripple of muttered profanities, but they seemed to express more amazement than outrage. Clearly, the men and women in the room had seen far worse in their day-to-day duties. But no one had ever seen a body in such an incongruous context.

The room grew quiet again as Rodriguez clicked to a close-up of the face and explained the importance of the disconcerting flies exploring its features. "Each stage of decay is characterized by a particular group of arthropods, each of which occupies a particular niche. Their activities are influenced by temperature, weather, and time of year." Though he struggled not to break stride, Rodriguez could not help but notice the commotion at the back of the room. Someone was shouting into the hallway. "Hey! Guys! Get in here. You've got to see this!" Others were literally pulling people in through the doorway.

Rodriguez described the four stages of decay he had observed and their direct relationship to the succession of insect families. Though he prided himself on having made his own insect identifications, he urged his colleagues to find qualified local entomologists to assist them in their own death investigations. Most important, he urged, "insect collections need to be part of the initial examination of human remains in the field." The identification of the insects may very well be the investigator's best hope of narrowing time since death, he concluded.

The thirty minutes allotted for the lecture had passed in a blur. As the lights switched back on, however, few people moved to leave. Instead, Rodriguez found himself mobbed. "How did you get away with it?" someone demanded. "This is amazing." "Unbelievable!" "How did you even think of trying to do something like this?" Best of all, Rodriguez heard someone say: "This is going to change the way everyone looks at time since death." The world of forensic science had been put on notice. Maggots were a homicide detective's best friend.

Unknown to anyone in the room that day, just three months earlier, in an eerily similar scenario in San Diego, California, another scientist had shocked a different audience with the same message and a slightly different slide show. Cutting from pictures of decomposing pigs to those of the Faith Hathaway murder, Meek presented his research to an audience of entomologists, explaining the potential service that they could provide to forensics. Admittedly, their response seemed muted in comparison to that received by Rodriguez in Orlando. Stunned might better describe the entomologists' reaction. Indeed, Meek's lecture room emptied far more quietly than it had filled up an hour before. Yet many of the scientists in the departing crowd that day left mesmerized by what they had seen, wondering how they might handle such a face-to-face encounter with death. In the coming years, nearly a dozen of them would find out.

7 The Dirty Dozen

WORMS'-MEAT, n. The finished product of which we are the raw material. The contents of the Taj Mahal, the Tombeau Napoleon, and the Granitarium.

—AMBROSE BIERCE (1842–1914)
THE DEVIL'S DICTIONARY

As Lee Goff settled into his seat for the eight-hour flight from San Diego to Honolulu, the images from the previous day's slide show continued to flash through his mind. A graying surfer-boy who'd traded his hangboard for a Harley, Goff had been struggling with the tedium of his work as an acarologist, or mite expert, at the century-old Bernice Pauahi Bishop Museum in Honolulu. Worse, the grant that funded his position paid less than Goff had earned as a graduate student. With a wife and two young daughters to support, he was perennially switching from one side job to another, most recently providing security for wholesale gift and jewelry shows. As a beefy Kenny Rogers look-alike given to motorcycle boots and denim jackets, Goff could look menacing enough. Nevertheless, he craved something that made better use of his Ph.D.

Meek's forensics presentation at the entomology conference in San Diego had not gone on long before Goff began to think that there might be something for him in this bugs-on-bodies business. The grisly side of the work posed no problem. Caught in the Vietnam draft fresh out of college, Goff had spent the good part of the mid-1960s as an orderly in an army morgue. It had not taken him long to differentiate the empty shell of the human body from its living essence.

Goff's superiors at the Bishop Museum did not warm to the idea. Founded by the nineteenth-century Hawaiian aristocrat Charles Bishop in honor of his wife, the last island princess, the museum had garnered a quiet reputation for research on the natural and cultural history of the Pacific. This was not the kind of publicity the museum needed, the administration insisted. It might even expose the institution to lawsuits. Goff found the attitude exasperating. Over the next year, he argued, largely in vain, that solving murder could hardly be termed "bad publicity." Meanwhile, one of the museum fly experts quietly slipped Goff the name of a kindred spirit on the mainland.

Paul Catts had also been in the audience during Meek's morbid presentation at the previous year's entomological meeting, and he too had left intrigued. A bot fly expert at Washington State University, Catts was already famous among students and faculty staff for the boyish enthusiasm that perennially sent his research in new directions. Not one to stay chained to a laboratory bench, he prowled the back roads of eastern Washington with students, pointing out the teeming carrion communities on assorted roadkill. On rare occasion, as he poked a flattened squirrel or tipped a bloated deer, Catts would actually see one of the livestock pests that were his specialty. But for the most part, he was simply enamored by the insect communities that materialized, as if out of nowhere, to feast on the fleeting repast. From the moment it hits the ground, a carcass becomes a fertile island waiting for the seeds of new life, he explained to anyone who would listen. But this island is destined to sink back into oblivion, unexploited, if the response is not quick. Hence, evolution has honed the senses of scavengers large and small to the exquisite scent of the dead. A blow fly, Catts pointed out with glee, could smell a dead squirrel over a mile away and follow the scent molecules like an airborne trail of crumbs, to reach the feast within minutes.

As for forensics, Catts's interest had been piqued years earlier, when as a professor at the University of Delaware he'd been asked to testify against a junkyard owner whose dogs had been impounded covered with maggot-infested sores. Catts had identified the fly larvae as a species that fed only on dead or dying tissue, meaning that the dogs' wounds had been festering, untended, long before the maggots took up residence. The dog owner was convicted of neglectful abuse of canines, and Catts came away fascinated with the stories bugs could tell.

When Goff called Catts's lab in the fall of 1982, the Washington State professor had his first homicide case in hand. Literally. Much to his colleagues' amusement, Catts had come to refer to the maggots in his rearing chambers—a row of empty tennis-ball tubes—by the name of the deceased. That was what they were, he insisted, intending no disrespect. "Little packets of Jane Doe." When they emerged from their pupae as winged adults, Catts would identify the species and, using Meek's protocol, count back to their mothers' arrival at the death scene. Catts and Goff hit it off immediately, the former giving the latter some much needed reassurance regarding their potential value to the world of forensics.

Goff resolved to make contact with local law enforcement. He decided to start with the Honolulu Medical Examiner's Office. After two months of unreturned calls, he finally badgered chief examiner Charles Odom into meeting him for lunch. Over a plate of curried rice, Goff tried to impress on Odom just how fast insects would respond to murder in the year-round warmth of the islands. Almost as fast, he promised, he could be at the scene with his collecting vials and net. Odom seemed receptive enough, and prospects began looking even better when, a few months later, Goff received a job offer teaching entomology at the University of Hawaii's College of Tropical Agriculture and Human Resources in the Manoa Valley outside Honolulu. Still fully intent on pursuing forensic entomology, Goff redefined both "agriculture" and "human resources" in his own mind and accepted the position. A professorship at the university would bring him within the network of state experts traditionally consulted by law enforcement, with the added bonus of graduate students to help him to begin the monumental task of cataloging the bugs that might show up on a real-life *Hawaii Five-O*. Clearly, the checklist of insects Meek rattled off during the slide show of his pig studies would prove little help half a world away from the Louisiana bayou.

Goff had yet to get a call from the Honolulu medical examiner when, in the spring of 1983, he read a newspaper account of a hiker's body discovered the day before beneath a treacherous, off-limits cliff in Yamea Falls State Park. When Goff saw that the hiker had been missing for over a week, he picked up the phone, dialed Odom, and hung on the line till he could get the pathologist to take his request to collect the insects before autopsy. Goff's arrival at the morgue sparked quizzical looks and obvious skepticism as Odom explained to the staff what the entomologist wanted.

"They were all standing around waiting to see if I'd throw up," Goff recalled later. But it wasn't disgust the professor had to hide as he tried to play the part of the seasoned professional. It was his childlike amazement at what he saw: Blow fly larvae, an assortment of second and third instars; and beetles—staphylinids and clerids, with their voracious fondness for the succulent maggots. "It was amazing to realize that I could actually recognize these things and why they were there," Goff remembers. "I actually knew what I was doing!"

Admittedly, Odom had no burning need for a time-of-death determination. With crushing injuries but no evidence of foul play, the young man's death was ruled accidental and dated to the day he failed to return from his hike. He had ventured off the marked trail and paid for his foolishness with his life. Nevertheless, Goff saw a chance to prove himself, at least to show that he could do his job without getting in the way. Quickly collecting his insect specimens in Styrofoam cups, he thanked Odom and left, pausing only at the secretary's desk for a cup of steaming water from the office teapot. To the woman's horror, Goff declined a tea bag and instead plunged half his maggots in the hot-water bath before adeptly transferring them to alcohol vials. Watching this shaggy hulk of a man stride out the door to the equally oversize motorcycle at the curb, a half-dozen staff members walked to the window, shaking their heads as Goff stashed his bugs under the seat of the Harley and roared away.

Back at his new lab at the university, Goff finished processing the evidence. The larvae that had been spared the earlier scalding he slipped into Styrofoam ice-cream cups for rearing into easily identified adults. He examined the preserved larvae under a dissecting scope to confirm their life stage, or instar, and make a preliminary identification as to species.

Looking over Goff's shoulder was Marianne Early. A scientific illustrator, Early had just finished the line drawings for the department's monograph on Hawaiian fruit flies, while scouting about for something to tempt her back into studenthood for another degree. She had met Goff months earlier, when he came to talk with her boss, dipterist Elmo Hardy, about the occasional forensic work Hardy was rumored to have done. Hardy handed Goff a thin file on the handful of fly identifications he had given the police over the past decade. "Here. This is what it's all about. I don't see the potential. But it's all yours."

What Hardy hadn't realized was that, along with his unwanted forensic work, he had given away his prize illustrator that day. Within months, Early was enrolled as Goff's first graduate student, with a master's project conducting decomposition studies on pig carcasses scattered across the island of Oahu. About the same time, Goff installed a police scanner alongside his insect incubators so he could keep his ear cocked for death-scene dispatches. Seventy percent of Hawaii's population and 90 percent of its murders were concentrated on Oahu, so Goff found there were few crime scenes his Harley couldn't reach within minutes, even when he forced himself to heed the speed limit. Getting past the yellow-tape police line was another matter. Goff was gaining a reputation among Honolulu police as a nutcase.

Finally, in the fall of 1984, Odom sent out the call. He needed help nailing down time of death for an apparent homicide. A local restaurant owner had been found dead in a drainage ditch behind an abandoned Pearl Harbor brewery. She had last been seen alive leaving her establishment with a tall white male weeks earlier. The medical examiner had no problem establishing her identity through dental records. But her body had decomposed so quickly in the tropical heat that he had no clue as to whether she died on the day of her disappearance or a week later. Knowing could make all the difference in whether the DA could make a case against the man with whom she was last seen.

Goff arrived at the morgue minutes later with Marianne Early on the back of his Harley. Because not much was left of the body, the pathologists had already completed their reports and quickly turned the autopsy room over to the entomologists. It took only a glance for Goff to understand their rush to leave. What the body lacked in tissues, it made up for in writhing and skittering bugs. Goff felt like rubbing his hands with glee.

It was exactly what he needed to test the computer program he had spent the previous month jury-rigging from business software designed to handle automated billing. Specifically, he had managed to modify its "quick-sort" function to handle a combination of insect species at specific growth stages, so that striking the enter key should extrapolate back to the beginning of their development—that is, time of death.

With this in mind, Goff and Early settled in to spend some time on the collections. From the back of the corpse, they collected three kinds of fly larvae: some chunky third instars of the flesh fly *(Sarcophaga spp.)*, as well as smaller larvae of bronzebottle flies *(Phaenicia cuprina)* and cheese skippers *(Piophila spp.)*, the latter named for its maggot's fondness for a certain fatty proteinaceous food and its remarkable ability to pop several feet into the air when disturbed. (Goff had to remove more than a few from his beard.) From the front of the body and the folds of a floral skirt, they collected hide and checkered beetles and the empty pupa cases of the hairy maggot blow fly, easily distinguished from other blow fly puparia by their rings of backward-pointing spines. Goff explained to Odom that it would take a couple of weeks for the larvae to grow into recognizable adults to confirm his identifications.

Two weeks later, Goff sat in front of his computer muttering quiet obscenities. The results suggested that either this person never existed or she died twice. Goff saw the problem. There was no way for the computer program to reconcile the presence of blow and flesh fly larvae on the back of the corpse and the empty fly pupa cases and beetles on the front. The body should have been covered either with maggots—suggesting a relatively short time since death, or empty pupa—suggesting a significantly longer time interval, enough for the flies to complete their development and depart. To have both on the same corpse made no sense.

Chagrined, Goff called the medical examiner's office and asked if someone could take him to the death scene. Perhaps he would find something there that could explain the contradiction. Not only did the medical examiner send a squad car, he came along for the ride, hoping to learn a thing or two. Their destination—the drainage ditch behind the abandoned Primo Brewery.

The stagnant water at the bottom of the ditch gave Goff a potential explanation for his problems. Odom confirmed that the woman had been found faceup and partially submerged in the water. That made sense to Goff, who imagined that the hairy maggot blow flies had moved in quickly to colonize the exposed upper surface of her body. Moreover, given this species' well-known aggressiveness, its maggots would most likely have devoured any other larvae attempting to invade their turf. Once the hairy maggots matured into flies and departed, beetles would have moved in to gnaw on the now-dried chest area.

Meanwhile, if the water level in the ditch had dropped even an inch, the less aggressive but more persistent flesh flies and blow flies could have found their way to the still moist skin on the underside of the body. Thanks to its previous submersion, the skin there would have remained tender enough to be penetrated by the maggots' delicate mouth hooks.

Goff returned to his lab newly confident he could make a determination. Discarding the developmental times of the flesh and blow flies, he concentrated on the hairy maggots. He knew from Early's decomposition studies with slaughtered pigs that, in Hawaii at least, hairy maggot flies generally arrived within minutes of death and spent up to six days laying eggs on the fresh remains. In the temperatures typical of late summer and early fall, it would take approximately eleven days for the eggs to hatch and the resulting maggots to mature, pupate, and hatch as winged adults. That told him the woman had been dead for a minimum of seventeen days. Add to this the cheese skippers, which prefer their bodies, like their dairy products, slightly aged. On Early's pigs, they generally arrived to lay their eggs four days or so after death, with their larvae reaching the size found on the victim's body around nineteen days. The presence of hide beetles, which hang back until the flesh is thoroughly dry, would likewise favor a slightly longer time estimate. Goff gave the medical examiner an estimated postmortem interval of nineteen days. It matched to the day the period of time between the discovery of the body and the woman's departure from her restaurant with the prime suspect.

Goff—or more accurately his cadre of insects—became the prosecution's prime witness, providing testimony at both the grand jury proceedings and the murder trial that followed. The accused drew a sentence of twenty-five years for second-degree murder, and Goff gained new respect from Hawaiian law enforcement. No longer the "nutcase," Goff became "the bug doc," welcome behind the yellow tape at all death scenes on the island.

* * *

AT THE SAME time Goff worked to master the necrophilous insects of Hawaii, back on the mainland interest in forensic entomology was multiplying. In the summer of 1983, Catts traveled to Walter Reed Medical Center in Washington, D.C., to help with the medical research of his former student and lasting friend, Wayne Lord. It was with Lord that Catts had begun his tradition of "braking for roadkill" on the streets and highways around the University of Delaware. Lord's research in the carrion communities of marine habitats had somehow metamorphed into a commission in the U.S. Air Force as a medical entomologist specializing in insect-borne tropical diseases.

But Lord's graduate research in decomposition continued to haunt him, benignly enough, in the form of periodic phone calls from some of the nation's leading medical examiners. As a doctoral student, he had given a brief presentation, "Decomposition in Northeastern Habitats," at a summer meeting of pathologists that included the likes of New York City medical examiner Michael Baden, best known for his investigation into the death of John F. Kennedy, and the famed criminologist Henry Lee, who created the University of New Haven's internationally renowned forensic science program in 1975. Both men continued to consult Lord, informally, on cases involving advanced decay, as did a growing number of pathologists who had heard of his expertise.

That summer, Catts and Lord struggled to keep their minds on their medical research. Repeatedly, their conversations diverged to forensics. Catts related the splattery details of his newest cases, then listened, transfixed, as Lord described the work of Bill Bass, the anthropologist who had enlisted him to write the entomology portions of student William Rodriguez's graduate exam. Lord had visited their unique field research station in eastern Tennessee. They were doing the unimaginable—observing carrion insects, not on wildlife carcasses, as he and Catts had done, but on actual human cadavers. The results of their research had been rippling like wildfire through forensic circles. Among the most keenly interested were the pathologists who'd been consulting with Lord in recent years. Several had approached him and Rodriguez about conducting lectures or workshops to explain more about these six-legged postmortem clocks.

Catts, in turn, shared his discovery of Finnish casework he found buried in a voluminous encyclopedia of forensic medicine. Admittedly, the work of Pekka Nuorteyva remained confined to the short Scandinavian summer when Nordic temperatures rose high enough to free insects from their "cold-blooded" torpor. Nonetheless, since the mid-1960s this blow fly specialist had amassed a casebook of murders in which he tested his ability to pinpoint time of death. Catts's favorite case involved the discovery of a decayed corpse in the isolated corner of a suburban park near Helsinki in early July. Someone had camouflaged the body with tree branches ripped from a nearby rowan tree. Police sent Nuorteyva two of the branches as well as soil samples from beneath the body, boxes full of chunky fly larvae, and a single brown puparium. From the soil, Nuorteyva extracted dozens of creamy white prepupae, their maggot bodies contracted into plugs, their skins not yet hardened and tanned into pupal cases. These he placed in a rearing can, left outdoors in a shaded spot similar to the ground where the body was found. On the rowan branches, Nuorteyva found a cluster of dried leaves that had been spun together by a colony of moth larvae before they wilted. A colleague at the University of Helsinki identified them as a species that spin leaves into nests in late May and early June. The size of the leaves, very close to full length, likewise suggested that the branches had been broken from the trees at the end of May or the beginning of June.

When the first flies emerged from the puparia in Nuorteyva's rearing cans on July 12, he identified them as Holarctic blow flies, *Phormia teraenovae.* Having reared the species under similar outdoor conditions, Nuorteyva was able to extrapolate back to the time of egg laying, drawing out his estimate to account for the cooler days of early June. The fly development must have started considerably earlier than midmonth, he concluded, most likely at the beginning of June. Independent police work confirmed that the man had been knifed on the second of the month, confirming Nuorteyva's deductions.

* * *

THE DISTRACTION FROM their tropical disease work only worsened that summer when Lord and Catts were joined by a lanky medical entomologist who had come to the military medical center to complete his annual six weeks of duty in the army reserve. As fate would have it, the reservist's father had literally written the book on North American blow flies. Rob Hall had never intended to follow his father's footsteps so closely. He had earned an English degree after an air force stint in the 1960s, with an eye on a career in agriculture journalism. Figuring that an added science degree would help his prospects, he counted on his childhood familiarity with blow flies to speed him through a master's program before entering journalism school. Hall had grown up with an insect net in hand, and by age six was already adept at finding the shiny metallic flies his father wanted, perched motionless at the tips of grass blades and stems. Later, when cataracts clouded the elder Hall's vision, Rob became his second set of eyes, describing the minute bristles and genitalia of the blow fly specimens beneath the lens of his father's dissecting scope. Only in this way could the Smithsonian's master dipterist continue to distinguish between many outwardly similar fly species.

After completing his master's degree, the younger Hall had allowed himself to be tempted into doctoral work by an entomology professor offering a paid assistantship. Though he graduated still intent on journalism, the final nail in the coffin came with an offer to teach medical entomology at the University of Missouri in 1978. Pushing thirty, with a family to support, Hall took the position and relinquished daydreams of scintillating exposés for the U.S. Department of Agriculture. He did not regret his choice, but remained watchful for something that might reengage his imagination.

In his own quiet way, Hall poured fuel on the fire of Lord and Catts's conversations that summer. He related his father's FBI casework with blow flies in the 1930s and 1940s, or at least his father's brief mention of it. A late-life baby, Hall had not come onto the scene until 1948. In particular, he recalled his father mentioning occasional visits from agents with vials of flies and maggots, invariably dead, dried out, and as a result often impossible to identify with certainty. His father gave little thought to the work, except for wondering what the agents did with the time estimates he gave them. Perhaps the information guided their investigations. But his visitors invariably disappeared as abruptly as they arrived. By the time the younger Hall was old enough to assist in fly identifications, his father had given up the bulk of his Smithsonian duties to someone with sharper eyes. Curtis Sabrosky, the Smithsonian curator of diptera from the 1950s well into the 1970s, would have been law enforcement's main point of contact regarding blow flies for nearly a quarter century, Hall figured. "He told me once that he did evaluations for the FBI on a regular basis. But he never mentioned what became of them."

But neither Smithsonian fly expert ever considered himself a forensic scientist or claimed to be able to pinpoint time since death, Hall noted. They based their crude estimates of maggot age on the developmental times of laboratory flies reared in cozy cages at constant temperature. Confronted with a fly for which they had no developmental data or nondescript fly larvae that defied identification, they would have been forced to extrapolate from a reasonably close cousin. For good reason, their opinions seldom if ever ended up as court testimony. "If challenged, they had nothing to show that their analyses reflected what actually happened on a dead body exposed to the elements."

To that end, Catts related a Nuorteyva case in which the Finnish entomologist placed liver bait in a sandpit where the bodies of two murdered hitchhikers had been found. In doing so, he attracted the same species collected from the girls' bodies. Leaving the rearing cans outdoors in similar environmental conditions, he then recorded how long it took the larvae to hatch and reach the size found on the dead girls. The resulting time estimate, substantiated by other evidence, led to a conviction.

If so-called forensic entomology was going to evolve from a quirky sideline into credible science, the three men agreed, it would require more field research such as Nuorteyva's in Finland, Lamar Meek's in Louisiana, and that at the anthropological facility in Tennessee. Catts, for one, vowed to start tossing out pigs when he got back to the Northeast. Clearly, he expected to see differences in species and growth times from those recorded in the Southeast. Lord resolved to step up his work educating pathologists on the importance of the insect clues that most of them still hosed from their autopsy tables. He also promised to keep his colleagues apprised of the insect work being done at the University of Tennessee through his friendship with Bass and Rodriguez. Bass, a master at explaining science to layperson juries, had already incorporated bugs into his court testimony to an impressive degree, said Lord. "Teenage maggots"—that's how he had described the fat third-instar larvae found on a two-week-dead murder victim. "We'd all do well to learn to communicate as well to a jury," Hall remarked.

The three men continued their conversations the next summer, by which time Catts had another murder under his belt and Lord had conducted a field seminar with Rodriguez, designed to teach homicide detectives, FBI agents, and medical examiners how to properly collect and preserve insect evidence at the scene of a crime. The response had been enthusiastic, with plans to expand the workshop the next year.

Meanwhile, at the University of Illinois in Chicago, forensic entomology's first minor celebrity was enjoying the attention of the journalists who'd been knocking on his laboratory door ever since his 1978 courtroom debut as an expert witness to a double homicide in the housing projects of East Chicago. Prosecutors had recognized a good thing when they brought their evidence—a handful of photos showing bug-infested bodies—to the distinguished-looking medical entomologist, whose specialty was the study of blow flies as carriers of polio, salmonella, and other infectious diseases.

Bernard Greenberg's baritone opinions exuded authority and cultured charm, and from the witness stand, he gently guided the jury through the essentials of blow fly biology, explaining how the insects in the crime-scene photos supported the prosecution's claim that the victims had been killed on the day that the accused was seen hefting several large, bloodstained bundles into the back of a rented U-Haul.

"Tale Told by a Fly," the first headline had read, followed by a story on the national AP wire that began, "Move over Sherlock Holmes . . ." Clearly, the headline writers enjoyed their new fodder, with spin-offs such as "The Strange Case of the Clues That Wriggle," "Super Sleuth Greenberg Uses Insects," and "Where Flies Flew Can Be Murder Clue."

The headlines, in turn, brought more forensic consulting work. Yet Greenberg never felt entirely comfortable with his time-of-death determinations, all of them extrapolated from the life cycles of his laboratory flies. He fully understood—and tried to convey to judge and jury—the many variables that could skew egg laying and larval growth on an actual corpse in a nonlaboratory environment. Then, in 1984, a case presented itself that both tested Greenberg's stomach for this kind of work and led to his development of a method for radically improving the accuracy of the blow fly as a postmortem timekeeper.

* * *

THE LAST THING Greenberg wanted to know about a case was when a suspect and victim were last seen together, or any other suggestion of the "when" of murder. He prided himself on his scientific objectivity and had no intention of giving a cross-examining attorney the chance to question his integrity, to suggest, in any way, that preconceptions or deliberate bias influenced his results.

But on June 19, 1984, there was no escaping the chilling details behind the box of maggots sitting on Greenberg's laboratory bench. The state police—in full panic at the possibility that they had a serial killer on the loose—had saturated the media with the details of child abduction and now, it would appear, murder.

On May 29, nine-year-old Vernita Wheat, of Kenosha, Wisconsin, had persuaded her mother to let her go on an afternoon ride with a friendly young man they had come to know over the previous month. "Robert" had promised to take Vernita to pick up a used stereo they described as a belated Mother's Day present. There would also be time to swing by a local carnival. A single parent, Vernita's mother was grateful for the attention Robert had been showering on her and her daughter. She agreed.

The pair never returned. The next day, Mrs. Wheat called the police in hysterics. She had already tried tracking them down at Robert's purported home address; it turned out to be an abandoned building. Down at the police station, she flipped through several notebooks of mug shots before she found him: Alton Coleman.

The story was not a reassuring one. Coleman's arrest record stretched back ten years, to a plea bargain that earned him two to six years in Joliet Prison for the abduction, rape, and robbery of a fifty-four-year-old woman. Since his parole in 1976, he'd wracked up another three rape charges, four of unlawful restraint, and a count each of deviant sexual assault and indecent liberties with a child. But nothing had stuck. Coleman's ability to terrify would-be witnesses was no doubt part of the problem, according to the judges and prosecutors involved. Police also suspected Coleman in the rape and strangling of a local teenager in 1982 but lacked the evidence to bring charges. Earlier in the year it looked like they finally had a witness willing to put him away, a Waukegan girl who said he had raped her at knifepoint. In fact, Coleman had been scheduled to appear at a pretrial hearing that very morning, May 30.

Later in the afternoon, police drove to the address on Coleman's trial papers and found his live-in girlfriend, Debra Brown. He'd been out all night, Brown admitted. He came home around 8 A.M., just long enough to change into his suit for the court appearance. He said he'd done something "real bad," but she hadn't seen him since. Police promised to return, warning Brown she needed to help them find Coleman before he got himself into worse trouble. She agreed. But that night Brown disappeared as well.

The next day, May 31, state police flooded local media outlets with Coleman's photo, details of the abduction, and notice of a $5,000 reward for information leading to his arrest. By the afternoon, a cab driver had called to say he picked up someone matching Coleman's description with a little girl just after midnight, May 29. He had dropped them near a scrap yard in Waukegan, Illinois. With evidence that the kidnapping had crossed state lines, the FBI issued a federal warrant and posted Coleman on its "Most Wanted" list. An FBI profiler warned that the psychological strain of Coleman's rape charges would be the kind of thing to trigger extreme violence.

The FBI's worst fears materialized on June 18, when seven-year-old Tamika Turks and nine-year-old May James of Gary, Indiana, disappeared on their way to a local store. That night, May was found close to death on a street near a wooded lot. In the hospital, she described how she and Tamika had followed a friendly man and woman who offered to show them some pretty clothes they had found in the woods. They'd strangled Tamika and left May for dead. She'd crawled to the street on her hands and knees after they'd left. When shown a photo lineup, May picked out Coleman and Brown.

The following day, police discovered what they suspected to be the decomposed remains of Vernita Wheat, lying beneath a dirty blanket in the bathroom of a boarded-up building near the scrapyard in Waukegan where she and Coleman were last seen. Nothing was left to identify the face. However, in the pockets of a green nylon jacket, the police found ticket stubs to a roller coaster and Ferris wheel. From the bones of her neck, they removed a wire television cable.

The forensic team tore down the entire washroom door and shipped it intact to the FBI laboratory in Washington for fingerprint analysis. From the tile floor, they scooped buckets of dead flies, brown pellets, and fat white plugs—maggots preparing to enter their dormant pupal state. They also caught a handful of strange spiderlike insects found skittering around the body. In the following hours, they would unfurl their wings to reveal themselves as newly emerged flies. Autopsy of the hollowed-out body revealed little except more maggots. The remains of more pale, spiderlike flies fell out of the girl's clothing, apparently crushed when the forensic team transferred the body to the morgue. The coroner immediately forwarded the insects to Greenberg.

At least, Greenberg mused as he unpacked his care package, they were now sending him more than *pictures* of maggots. Not so welcome were the next morning's headlines. FBI fingerprint experts had lifted a set of child-size handprints from the washroom door at the murder scene. They matched those taken from Vernita's schoolbooks. Worse, reports of yet another abduction had thrown the Midwest into full panic. The professor made sure his wife locked the door behind him when he left for the university after breakfast.

The larvae collected from Vernita Wheat's body completed their metamorphosis in Greenberg's climate-controlled laboratory, as Coleman and Brown continued their six-state Midwest crime spree. By the time the pair were apprehended on July 20, sitting on a bench in a suburban park north of Chicago, they had left a trail of at least seven bludgeoned bodies and as many brutally assaulted survivors.

Meanwhile, the FBI had managed to pull a smudged thumbprint matching Coleman's from the murder-scene door. Still, Assistant State Attorney Matthew Chancey knew the fingerprint did not prove that Coleman had been in the building on the night of the murder. It could have been left at any time. He needed solid evidence from Greenberg that Vernita was dead between the time she left with Coleman on May 29 and the time he arrived, alone, for his pretrial hearing on May 30.

The pressure only hardened Greenberg's resolve to conduct his calculations in such a painstakingly scientific manner as to defy challenge. All of the empty, pelletlike puparia were those of the black blow fly, *Phormia regina,* and so were the newly emerged flies around the body and the dead flies collected from the floor and windowsills. The young larvae collected at the scene likewise turned out to be black blow fly. Greenberg was not pleased.

He knew the black blow fly to have a relatively short life cycle of fourteen to seventeen days. So the offspring of the first black blow flies to find the body had clearly completed their cycle sometime before the victim was found. The younger larvae would represent a second generation. He could give a ballpark estimate as to when that second cycle began, but it would be just that, a guess. Greenberg could only hope that the intact pupae collected from the bathroom floor—now waiting to hatch in the rearing cages lining his shelves—would prove more helpful. He needed a longer-running clock—a fly with a life cycle that could span the interval between Vernita Wheat's disappearance and the discovery of her remains.

The ensuing wait gave Greenberg time to pursue an idea that had been percolating in his brain for months. Like any entomologist asked to estimate time since death from fly eggs, larvae, or pupae, Greenberg first identified the specimens and then consulted tables of how long each particular fly took to mature to a given stage under laboratory conditions. Since temperatures at a crime scene seldom if ever matched that of a climate-controlled rearing chamber, Greenberg had to then round up or round down his time estimate—often by several days if the temperature difference was large. Although he could easily justify this fudge factor to his scientific colleagues, it could be challenged in court. He could call it a scientific estimate. He could argue it was a reliably close approximation. Nevertheless, a cross-examining attorney could still call it a guess. "Conjecture, your honor. Move to strike."

What Greenberg had been contemplating was something *better,* something more objective, something mathematical. Agricultural entomologists called the concept "accumulated degree hours." It was, in essence, a measure of physiological time, and it enabled agriculturists to precisely calibrate their insecticide applications to coincide with a pest's most vulnerable life stage—for instance, the point when a wheat thrip molts from its first instar to its second instar form.

For over a century, entomologists had understood that outside temperature controls virtually every biochemical reaction inside an insect's body, including those speeding its growth and maturation. What the agriculturists discovered was that they could boil this temperature-dependent growth rate down to a simple formula of growth per unit of heat. In other words, if a given insect species takes 100 hours at 10 degrees Celsius (50 degrees Fahrenheit) to reach second-instar form, it will take fifty hours at 20 degrees Celsius (68 degrees Fahrenheit), or forty hours at 25 degrees Celsius (77 degrees Fahrenheit), or any other combination of hourly temperatures so long as they add up, in this hypothetical case, to 1,000 degrees Celsius. One had only to be careful that temperatures remained within the insect's minimum and maximum thresholds for growth. As a result, agriculture entomologists needed no more than a calculator and hourly weather station readings to tell local farmers when to spray their fields for maximum effect.

The method had never been used outside of pesticide application work, and then, only to look *forward* in time. In other words, to *predict* when an insect was about to pass from one life stage to another. Nevertheless, Greenberg saw no reason why the formula could not be flipped on its head and used to backtrack through earlier life stages until one arrived at the freshly laid egg. "It's only a matter of transposing the melody, so to speak, into a minor key," he mused.

Now all Greenberg needed was the right fly to emerge from the hundreds of brown pupae littering the floors of his rearing cages. Among the first arrivals were more black blow flies and a few dozen bronzy-green sheep blow flies, *Phaenicia sericata.* The latter, with only a slightly longer life cycle than the black blow fly, would not do the trick.

The one-month anniversary of Vernita Wheat's disappearance came and went, and the vast majority of the brown pupal cases in Greenberg's rearing cages had yet to open. The professor faced two distinct possibilities: Either the pellets contained dead, useless pupae or they would prove to be the perfect postmortem clocks for resolving this murder. On Sunday morning, June 30, Greenberg got his answer. Popping out of their pupal cases like sleepy campers from their zippered bags, they emerged pale and spiderlike. Over the next hour, their wings unfurled and their bodies darkened. As the room began to resonate with their raucous buzz, Greenberg saw his prayers answered.

Bluebottles, *Calliphora vicina*. With one of the longest development periods of any blow fly, the large, dusty bombardiers were exactly what was needed to span the interval between the girl's disappearance and the discovery of her decomposed body. From a known bluebottle developmental time of thirty-three days at 15 degrees Celsius (59 degrees Fahrenheit), Greenberg calculated a total of 11,880 accumulated degree hours to transform egg to adult fly. From the manila envelope sent to him by the DA's office, he pulled a readout of hourly temperatures requisitioned from an airport weather station a few miles from the murder scene. Tapping the figures into his calculator, he began subtracting each hour's temperature reading from the time the first bluebottle popped from its puparium in his lab. When the calculator stopped whirring, Greenberg arrived at midnight, Wednesday, May 30.

The chill that Greenberg felt as he stared at the calculator tape was not from visions of murder in a dark, boarded-up bathroom. It stemmed directly from the seductiveness of the "stopwatch" in his hand. The prosecutors would be thrilled with the "pinpoint" determination. Any jury would be impressed with its mathematical certainty. "Numbers don't lie." Isn't that what they say?

By contrast, Greenberg saw the potential for reckless abuse. It was one thing to tell farmers "exactly" when they should spray their fields, another to pretend to know exactly when a fly found a corpse. Unlike a certain nineteenth-century pathologist who found fame using temperature to pinpoint death to the minute, Greenberg had no illusions that he had done anything of the sort. True, he had accomplished exactly what he wanted—to increase accuracy and objectivity. What he feared was the use of the same method by people who would try to claim far more precision than it warranted.

In fact, the case before him clearly illustrated how any mathematical calculation of larval growth had to be combined with a thorough grounding in the habits of blow flies. Few if any flies would be buzzing about at midnight. Blow flies, a particularly light-sensitive family, instinctively settle when darkness falls. They might remain airborne around artificial lights. But the derelict building in which Vernita was found was shuttered and dark. The blow flies would have found her either in the waning light of May 30 or on the morning of May 31.

A longtime intimate of the blow fly's daily rhythm, Greenberg favored the dawn scenario. Once unleashed from their roosts by the first shafts of sunlight, the flies would have immediately followed the bewitching scent that had tantalized them through the night. Zigzagging through the air, they would have followed the trail of airborne molecules, its increasing concentration funneling them to their target.

Even so, it would have taken some time for the smell of death to filter out from a body wrapped in heavy blankets and closed inside an interior bathroom in a boarded-up building.

Greenberg double-checked the weather records for the morning hours of May 31. Temperatures had remained above 7 degrees Celsius (45 degrees Fahrenheit), well within the cold-tolerant bluebottle's active range. Taken together, Greenberg's impressive testimony would be enough to draw a conviction for murder in the first degree.

Among the first to hear of Greenberg's innovative use of accumulated degree hours was Lamar Meek, calling from Louisiana to invite Greenberg to give a presentation at the Entomological Society's 1984 Annual Meeting in San Antonio that November. So popular were Meek's previous talks that the conference committee had asked him to organize a half-day seminar. Paul Catts would be presenting some of his cases. Would Greenberg do so as well?

"Happily, Lamar."

* * *

NEAL HASKELL EASED himself into the front row of the San Antonio conference room with one hell of a hangover. An ice pack right below his receding hairline would have felt great. But Haskell thought better of the impression it would make on the professors in the room. "Hell, four months ago I didn't even know there was such a thing as a forensic entomologist," Haskell thought to himself. "Now I'm probably sitting in a room with every goddamn one of 'em." Before the day was out, Haskell intended every one of them to remember his name, that of the first entomology student to pursue a graduate degree in forensics. He had already obtained funding from the Indiana Coroners Association and had a good shot of getting more from the Indiana Medical School and the American Academy of Forensic Science. He'd even had his first case. Not as a forensic entomologist, per se, but as a deputized volunteer in the Jasper County sheriff's department. All the cops in Jasper County knew Haskell was the guy to call if you had bugs.

It was back in 1981 that the sheriff called, Haskell told the entomologist in the seat next to him, an interesting looking guy in red buckskin with a gash of a grin across his craggy face. "'Hey, Neal,' the sheriff said, 'we have a body down here with maggots on it.' Hell, I knew the guy . . . in better days, of course. God-awful how he'd drunk himself to death. But I could see right away by the maggots that he hadn't been gone long. They were small, first stage or early seconds. He'd been dead no more than four days, maybe five at the outside."

A bearded and boisterous Hemingway sort, Haskell had been out of college for decades when he paid a visit to his former entomology professors at Purdue University four months before the Entomology Society of America conference and became their oldest student. "I'd have been an entomologist long ago except my father died in 1966," he explained to the entomologist next to him at the conference. "Left me to run the family farm and take care of my mom." The man in the red buckskin nodded and extended his hand, "Paul Catts."

"Great to meet you, Paul!" Haskell clasped the professor's long bony fingers in his own beefy ones. He had found his first mentor.

The morning went by in a blur of thirty-minute presentations, from Greenberg's baritone narration of his meticulous developmental charts to Catts's wacky slide show of crime scene photos spaced by his own humorous line drawings of maggots in Sherlock Holmes garb and hapless police officers swinging butterfly nets over suitably skewered and fly-speckled victims.

When the bulk of the crowd cleared at noon, a dozen diehards remained behind to talk more murder and maggots. Someone knew of a decent place near the conference center where they could grab some lunch. After a few hours, the waitresses began leaving them alone and seating other customers away from their conversation. At 3 P.M. they were still there, Greenberg at one end, Haskell at the other. Between them, talking excitedly, sat Meek, Goff, Catts, Hall, Lord, Ralph Williams, Haskell's obliging faculty adviser at Purdue, and Ted Adkins, a veterinary entomologist from Clemson University on the verge of retirement and looking for something "to keep me out of my wife's hair." In the coming years, they would become the founding members of the Council of American Forensic Entomologists. Catts had already dubbed them the "Dirty Dozen"—as yet a few eggs short.

8 Perfecting the Postmortem Clock

These summer flies have blowne
me full of maggot ostentation.
—*LOVE'S LABOUR'S LOST* (ACT 5, SCENE 2)

IN THE DECADE since Meek's first murder case, his quirky sideline had become a bona fide science, albeit a science riddled with unanswered questions such as those being tackled by Gail Anderson of Simon Fraser University in Burnaby, British Columbia. Anderson had joined the ranks of the Dirty Dozen in 1988, and launched some of the most extensive field research they had ever conducted. Not only were she and her students documenting the arrival and development of insects on dead pigs in different environments across British Columbia, they were also refining the research to address the previously unknown effects of varying degrees of burial, water submersion, sun exposure, and scavenging, as well as the impact of apparel.

"In Canada, at least, our murder victims tend to be clothed," Anderson observed dryly when asked to explain the importance of the work. Early on, the work produced several important insights:

Clothing extends egg-laying time by holding in moisture and keeping skin tender and attractive to blow flies. For the same reason, it delays the arrival of insects such as ham and hide beetles that prefer their remains dry. Anderson and her students also noticed that maggot masses appeared sooner and grew much larger on clothed pigs, and so consumed them more quickly. Careful observation and reflection revealed why.

Most tiny, first-instar larvae must nurse on liquid protein before their needlelike mouth hooks grow large enough to tear through flesh. This is why blow flies concentrate their egg laying on open wounds and the moist mucous membranes found in the nose, mouth, and anal and genital areas. As clothing sops up blood and decay fluids, it becomes an additional nursery, multiplying the area open to egg laying and young larvae more than tenfold.

The Canadian research was progressing nicely in early 1995 when a waylaid e-mail threatened to scandalize the politically conservative western province that was home to the work. "Wanted, Pig Underwear," ran the headlines—first in a school bulletin and then a local paper, whose story was picked up by national and international news wires. "I thought I was going to get kicked out of the program for that one," recalled Leigh Dillon, the graduate student who had sent out the interdepartment plea for extra-large panties, bras, and men's skivvies. "There are some things you just can't find at Goodwill," she explained.

But if pigs in black net stockings and spike heels would further crime science, so be it. The Canadian federal government vowed to stand staunchly behind the work, and the Royal Canadian Mounted Police ensured that Anderson's field research remained the best funded anywhere in North America. Indeed, under Anderson's tutelage, in the early 1990s the Mounties added bug collecting to the standard investigative techniques taught to homicide detectives.

The decade since Meek first took the witness stand also drastically changed the treatment forensic entomologists faced in the courtroom. No longer could they count on taking opposing lawyers by surprise with their shocking testimony. Once a prosecutor's trump card, they now faced grueling cross-examination from defense attorneys who had learned to tap experts of their own. By 1995, the members of the Council of American Forensic Entomologists found themselves testifying almost as often for the defense as for the prosecution.

Overall, it was a positive change, though not a painless one. At times, equally qualified forensic entomologists came up with different estimates of postmortem interval based on the same insect evidence. They also knew how to punch holes in each other's testimony by exposing the soft underbelly of their own science, a science that still relied on more than a few untested assumptions.

It might be reasonable, for example, to assume that the developmental times of flies in different regions did not differ significantly. For that matter, it might be reasonable to assume that flies and maggots on corpses behave no differently than those raised on beef liver in a climate-controlled cage. Reasonable, perhaps, but open to legal challenge by an attorney who did not want an entomologist's damning testimony to be admitted into evidence.

Indeed, forensic entomology was coming of age at a time of unprecedented challenge to scientific testimony in the courtroom. Perhaps it was the trashing that forensics took during the O. J. Simpson trial. Clearly, judge and jury no longer took a scientist's word as gospel. The Supreme Court had also given lawyers a powerful tool for keeping scientific testimony not to their liking from ever being heard by a jury. In 1993, the Court's ruling in *Daubert v. Merrell Dow* drastically changed the rule of thumb that judges were required to use in determining whether to allow a scientific expert to take the stand. Previously, an attorney had only to show that the witness's methods were "generally accepted" as reliable by the scientific community. The new *Daubert* standard demanded far more. The test became whether the expert witness could show that his or her methods and conclusions had been *tested* under circumstances *directly applicable* to the case at trial. Forensic entomology had a lot to prove.

The good news was that word of the Dirty Dozen's outrageous field of study had drawn scores of captivated grad students—the underpaid foot soldiers any science needs to advance into uncharted territory. The bad news was the dearth of funding needed to pay even poverty-level stipends to these students, let alone cover the costs of publishing their research in peer-reviewed journals. Some professors, like Meek, found ways to tap funds for teaching assistants and divert dribs and drabs from general departmental budgets. Nevertheless, dedicated research demanded dedicated funds.

Helping matters was the splash made in law enforcement circles by the insect work at the University of Tennessee's Anthropological Research Facility. Neal Haskell became the first entomologist to benefit financially from that interest in the late 1980s, when his modest funding from the Indiana Coroners Association was followed by more money from the American Academy of Forensic Sciences and a large grant from the National Institute of Justice.

Haskell began by documenting what forensically important flies and other insects predominated in his home state of Indiana each season, the high and low temperature limits for their activity, when they got up in the morning, when their activity peaked, and when they went to bed at night. Haskell spent several nights camped out beside his pigs, checking for signs of nighttime fly activity. He did not see any, not even under a midsummer night's full moon. Yet this was no guarantee that "nocturnal oviposition" did not occur elsewhere.

In fact, the same summer Haskell spent camping out with dead pigs on his Indiana farm, Greenberg had a graduate student slipping out the back door at midnight to place preskinned and decapitated laboratory rats in her Chicago garden—some of them under the yellow glare of a nearby street lamp. Retrieving them before dawn, she brought the small carcasses to Greenberg's lab, where he slipped them into rearing cages to see what would grow out. Over the course of the summer, four out of the eleven midnight rats produced broods of three different forensically important flies—the bluebottle, *Calliphora vicina,* the sheep blow fly, *Phaenicia sericata,* and the black blow fly, *Phormia regina.* Admittedly, the nighttime carcasses did not produce nearly the volume of larvae as did bait placed out during the day at the same location. One carcass produced a single larva, whereas another yielded twenty, the latter being more than enough to influence a real-life murder inquest.

Had Greenberg's student accidentally placed her carcasses so close to night-grounded flies that they simply walked on board to lay their eggs? Or were blow flies in Chicago more willing to fly at night than those in Indiana? Insects were certainly capable of such subtle, regional quirks. In any case, such inconsistent results reemphasized the need for forensic entomologists across the continent to conduct field studies of their own. By the mid-1990s, they were tossing out dead pigs from the swamps of the Florida panhandle to the boreal forests of British Columbia.

At the same time, even the 50-pound pig that had become forensic entomology's standard proxy for murder was coming under courtroom attack. The entomologist could argue that the soft, near-hairless skin of a domestic pig closely duplicates that of a human, and that the torso of a luau-size porker parallels that of a 160-pound man. But there was *Daubert,* the new courtroom standard, demanding not just "general acceptance" but *directly* applicable science. "The case at hand is homicide, not pig slaughter," an attorney could rightly object. Even a judge unimpressed by such an argument might be loath to admit evidence that could provide the legal grounds for a conviction to be thrown out on appeal.

Recognizing what was at stake, the U.S. Department of Justice ponied up the funds for side-by-side comparative studies of pig carcasses and human cadavers. Clearly, there was only one place that such ghoulish research could be done. However, Bill Bass preferred that a trained entomologist be the one to do it. He invited Haskell to set up shop at his Knoxville research facility over the summer of 1988 and promised to supply the necessary human subjects.

Haskell did find some differences between the decomposition of his human and porcine subjects. After twelve days, his fifty-pound pigs were mere skeletons, but the human cadavers still had enough soft tissue to attract the interest of insects. Larger pigs might be better for studying the successive waves of different insects that colonize a corpse, he concluded. However, Haskell found no difference in the development time of the first maggots to appear on the scene—the blow fly progeny that serve as the entomologist's most dependable postmortem clock.

Meanwhile in Chicago, Bernard Greenberg continued to refine scientific understanding about the heat-fueled ticking speed of the tiny stopwatches. Since his introduction of the accumulated-degree-hour approach, the method had become widely used, both by qualified forensic entomologists and, as Greenberg had feared, by courtroom dilettantes. Worse, Greenberg saw problems that went beyond the technique's illusory to-the-hour precision.

At the core of the formula's simple arithmetic was the assumption that the speed of insect growth increased in lock step with outside temperatures. This proved true enough for individual insects, but as the tiny first-instar larvae of blow flies and flesh flies mature into chunkier second and third instars, as they settle down to the serious work of devouring a human corpse, they can turn into something else entirely. They can swarm. The resulting activity becomes not so much that of individual maggots, but that of an all-consuming pack. The teeming mass churns and roils within the cadaver, with thousands of maggots diving for food, then rolling to the surface for air and plunging down again. The maggot mass becomes an ecosystem unto itself. It becomes the source of the ghoulish steam that has risen from cold battlefields since the beginning of man's inhumanity to man. The resulting heat—whether from the friction of their roiling movements or the combined chemical spark of ten thousand tiny, flesh-filled guts—can sustain larval growth even in subfreezing weather.

This heat generation, Greenberg concluded, must be taken into consideration when calculating temperature-dependent growth rates in the later stages of larval growth. How could one quantify such an effect? The experiment he set up to answer the question was pure Greenberg—exacting, laboratory-controlled, and as meticulously clean as one can get studying decomposition and maggots.

Not one to muck around with dead pigs, Greenberg flipped through the pages of an anatomical supply catalog until he found what he needed: a scrubbed, sanitized, and conveniently hinged human skull. Greenberg stuffed the brain case with raw ground beef and placed it in a mesh cage filled with black blow flies, *Phormia regina*. By the end of the day, the flies had plastered the skull and burger with some 10,000 eggs.

Over the next week, Greenberg tracked temperatures within the skull with a digital thermometer stuck through the orbit of one eye. Over the first few days, the tiny first-instar larvae had little thermal effect. However, as they grew into chunkier second and third instars, the heat within the skull began rising, peaking at about 18 degrees Celsius (10 degrees Fahrenheit) above room temperature just before the third instars reached their maximum size. The maggot mass temperature quickly fell off again as soon as the postfeeding larvae began abandoning the hamburger brain to seek hiding places for pupation.

How much more heat might be generated by the massive maggot masses found roiling through a decomposing human torso? Greenberg wondered. Indeed, the characteristic roiling motion of a large maggot mass appeared to be a temperature-regulating mechanism: Overheated maggots at the core of the mass continually churned to the surface to cool before plunging down to the feast again.

Greenberg demonstrated yet another source of unaccounted heat with rabbit carcasses placed outdoors in sun and shade. Greenberg found that the skin and internal temperature of the sun-drenched carcasses averaged 30 to 50 percent higher than that of those lying in shade. Moreover, the sun-warmed bodies retained their heat into the night, when outside temperatures dropped to inhospitable levels. No doubt, a larger, human cadaver would retain more solar heat, most dramatically if it happened to be dressed in dark clothing. Yet an entomologist who calculated larvae development time solely by temperature records from nearby weather stations would miss these differences entirely.

In yet another experiment, Greenberg found that blow flies reared at the constant temperatures typical of laboratory rearings do so at a quicker rate than flies under more natural, fluctuating temperatures. His finding directly challenged the simplicity of his earlier accumulated-degree-hour approach, which calculated a maggot's age as a simple sum of its thermal history. In 1991, Greenberg reported his cautionary results in the *Journal of Medical Entomology*. His fellow forensic entomologists took note, and began sharing their own doubts and cautions.

What became obvious was that forensic entomologists needed more than a box of bugs and a weather report if they were to refine their estimates of time since death. They needed specific information about the microclimate where the corpse was found. Did it lay in sun or shade? On a windy hill or in a protected cove? Had there been a maggot mass present? If so, what was the temperature in the middle of the mass?

In the best of all worlds, the entomologist would record such information him- or herself, as part of a forensic team responding to any death scene. But outside the beeper-wearing Goff, who cruised Oahu with collecting vials and a butterfly net under the seat of his Harley, few forensic entomologists had the opportunity to do so. By contrast, the rest of the Dirty Dozen had forty-nine states and ten provinces to cover. They had to rely on detectives, crime technicians, and coroners to be their eyes and hands at the scene of a crime.

Their hope of improving their time estimates lay in educating these partners, not just about the importance of collecting bugs, but also in the finer details of *how* to do it—with proper technique, thorough observation, and clear notation. Their educational outreach efforts swung into high gear in 1988, in the back room of a Vancouver microbrewery in the waning afternoon hours after yet another standing-room-only forensics symposium, this time at the 18th International Congress of Entomology.

In the four years since the group's last major symposium, Haskell had become the field's first full-time consultant, with a farmhouse laboratory and a mobile response unit—a rusting Dodge Colt sporting the Indiana license plate "MAGGOT." Lord had been hired on as a special agent within the FBI's serial killer unit. The rest of the Dirty Dozen had a flush of graduate students assigned to both field research, to record the day-to-day succession of insects on a carcass, and laboratory work, rearing flies in controlled conditions to document their developmental times. Equally vital were ecological studies such as graduate student Jeffrey Wells's study of the encroachment of the hairy maggot blow fly, *Chrysomya rufifacies*. An exotic species moving into North America from Asia and points south, the fly was no arcane matter. The notoriously aggressive maggot had an appetite for its smoother cousins that could reset a postmortem insect clock by clearing a corpse of all earlier arrivers.

Attending the group's overflow program in Vancouver that year was Finland's esteemed Nuorteyva, come to share his cases and learn about the American research that was bringing to fruition ambitions born in Europe over a century earlier. More than a few rounds of beer followed at the local microbrewery.

Having cleared the surrounding tables with their usual conversation, ten entomologists settled down to discuss a project that most of them felt was long overdue: a North American field manual, as free as possible of scientific jargon, to guide both homicide investigators and nonforensic entomologists through the collection of insect evidence and the notation of environmental conditions crucial to refining any estimate of time since death.

The manual would also serve as a primer for hands-on workshops, some of the largest of which Lord taught on the wooded campus of the FBI Academy in Quantico, Virginia. Haskell, too, had begun taking his show on the road, packing his rusting "maggotmobile" with supplies and driving a circuit that extended from coast to coast, Ontario to south Texas.

Divvying up the chapters, the group hammered out *Entomology & Death: A Procedural Guide,* in just over two years, dedicating it to the "luckless victims of homicides whose plight has made this endeavor necessary" and, on an intensely personal note, to their own Ted Adkins, who had died of heart failure before the work was complete and whose widow, Joyce, had stepped in to publish the work out of her South Carolina print shop.

The contents of the 200-page manual dramatized just how far the science of forensic entomology had come in the half century since Hoover's G-men began bringing the occasional envelope of flies to Rob Hall's father. The list of collecting materials and equipment alone spanned two pages, from electronic thermometers and psychrometers (for measuring humidity) to light-tension larval forceps and insect aspirators (for sucking up the smallest specimens) to ten kinds of preserving solutions (with recipes found elsewhere in the book). Page after page of forms and instructions detailed the procedures needed to process crime-scene "fauna," beginning with visual notations taken before so much as a shadow has touched the body. Among the most crucial observations: the location of major insect infestations (often a sign of underlying trauma), obvious size differences in larvae (a sign that they may have come from eggs laid at different times or by different species), and signs of maggot predators such as beetles and fire ants that could have depleted maggot numbers or delayed their establishment.

To determine the microclimate in which various insects developed, the entomologist also needed a description of the body's position, including notations of which body parts lay in sunlight and which in shade, which came in contact with the ground, even the compass direction of the torso, the latter an aid to finding the postfeeding maggots that often abandon in a southeast direction, as if to anticipate the warmth of the morning sun.

Courtroom challenges had also taught the entomologists to ask for temperature and weather data, not only from the nearest weather station but also from the murder scene itself. In this way, they could look for and adjust any differences between the ambient and ground temperature around the corpse and that from the weather station, whose readings enabled them to extrapolate back over the previous days, weeks, or months.

Equally crucial in the presence of a maggot mass was a temperature reading taken from its midst. Fortunately, the same temperature probes used by coroners to gauge cadaver temperature could be thrust into the center of a maggot mass, then held over the corpse to measure air temperature, and finally slipped beneath the body and the ground, a favorite breeding site and hideaway for dozens of forensically important species.

The entomologists also needed appropriately labeled samples of insect specimens collected from each site of infestation, including a pinch of each size of larvae to be seen. Most crucial were the specimens that represented the first generation to develop on the corpse. Often, these stood out as the largest maggots on the body. However, they might be slightly smaller, "postfeeding" larvae already inching away from the scene to pupate in the soil or some dark, indoor crevice. For that matter, they might already be hidden away in the midst of pupation or already hatched, leaving only their papery pupal cases as evidence of their emergence. Consequently, a proper collection was not complete until someone had thoroughly spaded the soil under, adjacent to, and up to three feet away from the body, or checked under every carpet and sofa in the room.

For all this, no task proved more vexing for the typical homicide detective than the wielding of the proverbial "butterfly" net. In addition to the raucous laughter the act invariably provoked in fellow officers, the task proved deceptively perverse. More often than not, a sweep through a veritable cloud of flies served only to scatter the agile, strong-flying insects. Be it at a field workshop or actual crime scene, nothing elicited profanities quite so predictably.

But the collection of adult flies could prove crucial in determining time of death when a corpse bore no more than a smear of eggs or the tiny first instars that defied even the most skilled entomologist's identification. Nor did the fragile eggs and first instars always grow out in the entomologist's lab. They often failed to do so, even after meticulous collection and prompt delivery. In such cases, adult females netted at the scene provided the crucial species identification needed to guide the scientist's time estimate.

That said, few crime technicians, coroners, or detectives had the patience to learn the unhesitating sweep and artful twist required to trap a cloud of blow flies in one fell swoop, ending with the long tail of the net neatly flipped over the opening ring. Taking pity, Haskell ultimately designed a kind of pup-tent sticky trap that took advantage of the female blow fly's habit of periodically leaving a corpse to rest in nearby vegetation. Set near the body on clothespin legs, the trap soon filled with flies and could be gently folded, labeled, and mailed to the laboratory of the consulting entomologist.

Entomology & Death was also a casebook, with descriptions of a decade's worth of actual death investigations, judiciously stripped of identifiers such as names and dates. Each case involved an entomological estimate of postmortem interval that had been subsequently shown accurate by confession or other evidence. This was, after all, the proving ground for the entomologist's worth to law enforcement. Although science demanded experimental validation, years of throwing out pigs and raising countless legions of flies in the laboratory meant nothing unless it translated into stories of "true crime" success.

Casework was also the entomologist's only chance to put scientific dogma to the test in the uncontrolled and unpredictable hell of real-life homicide and unexplained death. Just like the forensic pathologists and forensic anthropologists who came before them, the entomologists were very much learning by the seat of their pants. As a result, their lectures and café gab sessions became crucial forums for swapping teaching stories such as Paul Catts's intriguing Case of the Massive Maggot.

* * *

IN MID-OCTOBER 1987, Catts had been asked to help date the death of a young woman found stabbed in the chest, facedown and shoeless, near a logging road in the pine woods northeast of Spokane, Washington. From the putrefaction that had already blackened the face and marbled the extremities, the medical examiner could only guess she had been dead days to weeks. For a closer estimate, he sent Catts an abundance of maggots collected from the woman's face and chest wounds. Half of them Catts immediately preserved in an alcohol solution. The other half he grew out on beef kidney until they emerged, twenty-one days later, as a shimmering cloud of bronzy-green and deep metallic blue bombardiers that Catts immediately recognized as sheep blow flies, *Phaenicia sericata,* and shiny bluebottles, *Cynomyopsis cadaverina.* He knew that both species begin laying eggs within the first day of death.

Once identified in this way, the maggots preserved on receipt provided Catts with the stopwatches he needed to count back in time from the discovery of the body. They fell into three distinct size classes. The smallest would have been the subsequent broods of flies that lingered at the scene to mate and lay eggs again over the first few days after death, before the body grew unpalatable to their discerning taste. The larger maggots—at least all but one of them—matched the size of seven-day-old larvae that Catts had previously collected from experimental pig carcasses decomposing in virtually identical weather and habitat. Had it not been for a single, mismatched maggot, Catts would have picked up the phone to give the coroner his one-week time estimate. However, the solitary maggot was impossible to ignore. At nearly three-fourths of an inch (18 millimeters), it was massive, even for a well-fed, third-instar larva. Such a length would suggest a time interval between death and discovery of at least three weeks, pushing the stabbing back into mid- to late September. Impossible, the DA told Catts. The young woman, who had since been identified from clothing and dental records, was seen alive during the first week in October.

Could the maggot have crawled onto the corpse, say, from a nearby animal carcass? Catts asked investigators to revisit the scene to check the remote possibility. They found no such carrion.

It was only when Catts shifted his gaze from the fat maggot floating in the preserving solution to the specimen label on the vial that an alternative explanation presented itself. "From naso-oral area" read the collecting details. Over lunch that day, Catts asked insect physiologist and office mate John Brown whether the maggot's growth could have been enhanced if it had developed in, say, some cozy, cocaine-laced pocket in the victim's nose.

"For god's sake, Paul, could you just once wait until after I've eaten," Brown groused, before assuring his colleague that, yes, such a scenario was entirely possible.

Catts called the DA that afternoon. Did the victim have a history of cocaine abuse? Subsequent questioning of witnesses revealed that she had, and that she had in fact been seen snorting immediately before leaving with the man accused of her murder—one week before her body was found.

* * *

OVER THE ENSUING years, Catts, Lord, Goff, and others would work a number of cases and conduct laboratory research to confirm that a variety of drugs, from cocaine in nasal tissues to tricyclic depressants permeating the entire body, could dramatically accelerate or slow insect development. An added bonus of their research was the discovery that flesh-eating insects concentrate these drug residues both in their flesh and in their hard pupal cases, both of which can be used as toxicology samples long after the corpse itself has decomposed beyond testing.

Casework likewise taught the entomologists they must factor "access" into their time-of-death determinations. Closed windows and doors posed formidable but not always impenetrable barriers for death's most avid fans. Flies, in particular, had a way of relentlessly wriggling through the smallest cracks to get to their goal. In one of Greenberg's death investigations, flies gained access to a pair of bodies in a closed car trunk via a small drainage hole in the spare tire well. In another, flies finagled their way into a closed and darkened house despite weather-stripping on the doors and windows. Then there were your standard bodies-in-garbage-bags. Even if one could design an experiment to time how long it takes a blow fly to wrestle its way through a given gauntlet of murder-scene obstacles, how could anyone factor in how tightly different people twist that trash-bag tie? But every now and then, a homicide begged for just such an exacting replication.

Goff often told of such a case, involving a wrapped and sealed bundle discovered by the residents of an upscale suburb on the North Shore of Oahu, a bundle they had good reason to believe might contain the body of a neighbor who'd gone missing a couple weeks before. When Goff pulled up to the scene in the fading light of a December afternoon, he saw that police had yellow-taped a dense thicket of mimosa trees running alongside an access road separating a golf course and an area of widely spaced exclusive homes. An impromptu neighborhood search party had turned up the long oblong package—a rolled blanket sealed with elastic bandages—after someone mentioned that the missing woman's estranged husband had been seen prowling there. They had unsealed a corner just enough to get a glimpse of a decomposing hand before calling the Honolulu police.

Having come to respect their resident "bug doc," the officers had left the blanket roll intact for Goff's inspection. From the outside surface, he removed a pasting of fly eggs. Peeling off the elastic bandage and unfolding the tucked ends, he found a second, inner layer of blanket peppered with hister beetles, *Saprinus lugens,* feeding on third-instar fly larvae, and an abundance of unopened pupal cases. The pupa's distinctive, backward-pointing spines immediately identified it as that of a hairy maggot blow fly, *C. rufifacies.* From the body itself, Goff collected more unopened pupae and a variety of third instars, both hairy and smooth. From soil samples spaded from where the bundle had lain, he found nothing. It would appear the tight seal had prevented the fattened maggots from making their usual exit.

Two days later, the flies began popping out of their pupal cases in Goff's lab. Calculating degree hours from a nearby weather station, he counted back to egg laying approximately ten days, twelve hours, before the neighbors found the body.

The woman's estranged husband came to trial for murder some months later. Even more damning than the insect evidence were the recollections of a next-door neighbor who had seen the woman's estranged husband enter the home thirteen days before the body was found. She had also heard what, with twenty-twenty hindsight, she took to have been sounds of a struggle. The defense focused on the two-and-a-half-day discrepancy between the witness's incriminating story and the time of death estimated by the insects. When Goff suggested that it might take that long for the blow flies to have wriggled their way through the wrappings to the body, the defense pressed him to be more exact.

Purchasing blankets, rolls of elastic bandage, and a fifty-pound pig from an island slaughterhouse, he duplicated the murder in the overgrown brush of his own backyard. Every four hours, Goff covered both himself and the carcass with mosquito netting to check whether flies had gained access, then carefully rewrapped the pig for the next watch. Over the next two weeks, "things got a bit twitchy with the neighbors," he recalls. Nevertheless, Goff got his time delay: two and a half days.

Yet another of Goff's stories illustrated that barriers to insect access are not always so obvious. In his delightfully droll style, Goff describes the lesson in his autobiographical *A Fly for the Prosecution*. A golfer on one of Hawaii's many popular golf courses had hit his ball into the rough near the sixteenth hole, where he found an apparent suicide victim still hanging from a tree branch. The amazingly determined golfer finished his game before reporting the body to the police. When Goff arrived on the scene, he found himself puzzled by the highly unusual pattern of insect infestation on the corpse. It bore far less than the usual number and variety of insects for its advanced state of decay. In particular, Goff noticed that many of the beetle species that *should* have been preying on maggots or gnawing on the dried skin were either wholly absent or present in unusually low numbers only around the man's feet. Later, when the body was identified, Goff figured out why.

Much to the homicide investigator's surprise, the victim turned out to be a man who in life had been a diminutive five foot two inches. By contrast, the decomposing corpse had played out to nearly six feet by the time it was discovered with its toes just brushing the ground. Goff surmised that the crawling beetles simply could not gain access to the body until it had elongated enough to make contact with the soil where they had gathered. The lack of contact with terra firma might also explain the unusually low numbers of blow fly larvae Goff collected. At the height of maggot-mass frenzy, larvae continuously tumble off the corpse as they circulate to the outside of the mass to cool. Normally, they simply clamber back aboard and dive in again, but so long as the hanging victim's feet remained out of reach, they could not.

Goff had the opportunity to test his theory when invited to coteach Wayne Lord's annual forensic entomology workshop at the FBI's Quantico campus the next year. Sure enough, a hanging pig gathered few maggots . . . and no crawling beetles. However, in the case of the test pig, the drip zone directly beneath the oozing carcass provided ample food for maggots to continue their development on the ground. Thinking back to the same area beneath the hanging victim on the golf course, Goff realized why he had not seen similar activity. "The golfer who discovered the body, apparently upset by the sight of the corpse, had obviously needed several swings to hit his ball out from under it, and the contents of the drip zone were scattered across the course."

In subsequent years, Goff tested out similar questions about insect-activity delays. It would be reasonable to assume, for instance, that burning—another depressingly common murder scenario—could delay the start of death's insect stopwatch. When one of Goff's graduate students, Frank Avila, simulated such an arson-homicide, he saw adult flies landing on one part of the pig carcass even before the flames on another part had stopped burning. If anything, burning appeared to make corpses more attractive to blow flies, though Avila could not fathom the biochemistry that might explain why.

Even more informative was a human immolation investigated by Bill Bass. Although Bass did not particularly like testifying as a bug expert—he was, after all, an anthropologist, not an entomologist by training—the insects he encountered in his work continued to play an important role in his time-of-death determinations. This particular case involved a body in the backseat of a burned-out vehicle discovered by dirt bikers riding through an off-road canyon. When Bass stuck his head through the car's back window, he saw immediately why the police had called him—the victim had been burned beyond superficial recognition. Given the low flammability of the synthetic upholstery, it was a sure sign of deliberate burning with accelerants. Bass also noted an abundance of small maggots on the charred surface of the body. He eyeballed them at roughly two to three days old, then quickly moved to a more compelling observation.

The fact that the victim's skull remained intact spoke volumes to the veteran bone detective. He knew from long experience that flames hot enough to remove all skin from the face, neck, and chest—as they had done here—would have also set the fluids inside the skull to boiling. Unless there was a steam outlet to relieve the pressure, the brain case would have shattered into so many quarter-size pieces. Back in his laboratory beneath the stands of UT's Neyland Stadium, Bass cleaned the skull of its remaining charred tissue to search for that steam outlet. His thumb found it, a perfectly round gunshot wound at the back of the head, low down and just above where the spinal cord would enter the brain. What happened next would remain forever etched in Bass's memory. As the anthropologist turned the skull around in his hands, the brain case crumbled. Looking inside, he saw, not the usual amorphous mass of coagulated brain tissue, but what he judged to be about a hundred plump, cooked maggots. From his reference collection, Bass judged them a little over two weeks old. Teenage maggots, just about to pack up shop and turn into adult flies.

Putting it all together, Bass saw a vivid picture of an execution-style murder sixteen to eighteen days before the body's discovery. The first flies on the scene had simply followed the bullet into the brain. Two weeks later, someone got worried and went back to burn the evidence. As soon as the fire went out, the flies returned, their second-generation larvae clocking another two days of growth. Bass's backtracking led police to the missing person report filed by the dead man's employer. The anthropologist's bone measurements confirmed the identification. Unfortunately, police never found the killer or killers.

For the most part, forensic entomology's role in time-of-death determinations continued to play this sort of supportive role—guiding the usual search through missing-person reports or bolstering the accounts of those who'd last seen the victim alive. On occasion, however, death's tiniest stopwatches took center stage, unsupported by any other shred of evidence.

In the summer of 1990, the Tacoma police found a thirty-four-year-old man dead in his apartment. At first it looked like a fight—blood splattered around the bedroom, mattress on the floor, a badly decomposed body tangled in the sheets. However, outside of a small crack in a bathroom window, the apartment was sealed tight, every door and window frame latched from the inside. Autopsy found a single gunshot wound through the neck, sending police back to the scene for an exhaustive search for the bullet. It finally turned up in the wall behind the man's bed, hidden by a picture frame that had been knocked off kilter in the apparent struggle. The bullet did not match the only gun found at the scene, an unused handgun in the drawer of the dead man's bedside table.

The bewildered police sent Paul Catts the only other evidence they had: boxes of papery pupal cases and dead flies and a smattering of maggots. Catts figured he had two generations of flies on his hands and, given temperatures for the time in question, came up with an estimated time since death of six weeks.

When local police reviewed their logs, they came up with a spate of noise complaints about a raucous party across the street that matched the date and locale. Sometime after midnight, a particularly obnoxious reveler had begun firing shots in the air. When the police tracked him down, they found that the bullet in the wall matched his gun precisely. Ballistics experts then retraced the path from the party to the man's bed. The bullet had taken a particularly bizarre course, ricocheting off the metal beams of an adjacent garage before entering the bedroom and passing through the victim's neck. Clearly, death had not been instantaneous, given the apparent signs of struggle.

Other times, an entomologist's time-of-death determination spoke to the sheer heinousness of a crime. Neal Haskell remembers one such case that continues to haunt him. On April 14, 1996, police found four bodies inside a cabin, locked from the outside, in New Mexico's Manzano Mountains. In one bedroom lay a known Albuquerque gang member and his girlfriend, apparently shot in their sleep. More disturbing, police found the girlfriend's sons, ages three and four, dead in the next room. Within days, police arrested four members of the dead man's gang. Two of the suspects accused the other two of the murders, and those two claimed alibis for the previous month, during which time witnesses reported hearing music and voices coming from the mountain cabin. Police sent Haskell photos of the crime scene with hopes of showing that the victims had been killed earlier than that, for they had not been seen by their families since early December.

From the photos, Haskell could readily see thousands of dead flies and pupa cases on the floor and windowsill of the bedroom where the gang member and his girlfriend had been shot. New Mexico entomologist Pete LaScala returned to the mountain cabin to collect a sampling of the dead flies, which he identified for Haskell as common bluebottles, *Calliphora vicina.* Checking local weather records, Haskell found a single warm spell in mid-January that could have stirred the flies from their winter torpor. The man and woman had to have been dead and available for colonization by that time, considerably before the period for which the suspects had alibis.

But the photographs of the second bedroom would prove far more disturbing, given what Haskell knew about bluebottle development. In contrast to the adults' room, the preschoolers' room appeared relatively free of dead flies. Instead, Haskell saw just a smattering of live maggots on the face of the older brother. Much to his horror, Haskell could not escape the only entomological explanation for the difference between the two bedrooms. The maggots in the boys' room were not from the same generation of flies as those next door. Before the children died, the maggots in the next room had to have completed a full life cycle and given rise to a second generation of flies, a few of which survived long enough to colonize the boys. Because bluebottles require at least 300 accumulated degree hours to complete their development, they told Haskell that the children had remained alive and locked in the cabin for weeks after the murder of their mother and her boyfriend.

The pathologist's report confirmed Haskell's horrifying scenario. The only obvious injuries found on the boys had been torn and bloodied fingertips, most likely from trying to break through the heavy screens nailed across the cabin windows. They had died of thirst and starvation.

Given the callous cruelty of the children's deaths, the prosecution sought the death penalty in each of the four separate trials of the accused gang members. Nonetheless, with only the suspects' conflicting accusations tying any of them to the crime, only one of the trials ended in a conviction, with a sentence of more than 130 years in prison.

Hundreds of such cases served to cement entomology's place in standard homicide-investigation procedures during the 1990s. The credentials of an individual entomologist could still be challenged in court. However, the science itself had proven its worth beyond reasonable doubt. Indeed, by the second half of the decade, *failure* to collect insect specimens became a potential challenge to the sufficiency of an investigation. A defense attorney could rightly claim that evidence capable of supporting her client's alibi—that he was elsewhere at the time of death—had been carelessly disregarded.

For the most part, however, death investigators looked to forensic entomology only in the extended postmortem period. That is, after the pathologist's triple time markers of body heat, rigor mortis, and lividity had faded away. Like the forensic anthropologists who preceded them, the bug docs inherited the medical examiner's hopeless cases. But for those who thought to ask, there were times when insects, or the lack of them, could measure minutes since death with more certainty than any autopsy room yardstick. The pages of Haskell's gruesome casebook cough up another illustration.

Haskell got the call in his farmhouse laboratory in April 1992. Two members of Miami's notoriously violent Yahweh Ben Yahweh cult had cut a deal. They were getting off the hook for drug-possession charges in exchange for their testimony that former cult member Richard Foster was responsible for the unsolved 1981 murder of another cult defector.

Foster's public defender was now calling Haskell to ask his help in exonerating his client. "I'll call it as I see it," is all Haskell would promise. The defense attorney recounted what he knew: Around 10 A.M., the morning of November 13, 1981, a wildlife ranger patrolling the Everglades west of Miami spotted a blanket on the side of the road that he was sure had not been there when he'd passed by a couple of hours earlier. Climbing out of his truck to investigate, the ranger pulled back the blanket to discover a headless body, still spurting blood from the neck. The terrified ranger, fearing that the murderers might be near, jumped into his truck and sped back to his station to call the police.

Nearly eleven years later, Foster stood accused of the murder, despite the fact that he had been at work, surrounded by witnesses, the morning of the body's discovery. To make charges stick, the prosecution contended that the ranger's account of "spurting blood" was unreliable, the exaggeration of a hysterical witness. By contrast, the prosecution's plea-bargaining witnesses claimed that Foster had beheaded the victim with a machete the previous afternoon. Could Haskell disprove their "lies?" the defense attorney wanted to know. The autopsy report had proven useless in confirming or refuting either alleged time of death.

Haskell was intrigued and agreed to take the case, asking the attorney to send him copies of police reports and weather records. Temperatures had remained in the 70s and 80s throughout the period in question. Moreover, the crime technicians who had processed the death scene understood the importance of insect evidence. In this case, there had not been any to collect. They noted a few flies hovering near the remains, but found neither eggs nor larvae on the body itself. The autopsy report likewise noted an absence of insects.

When called as a defense witness at Foster's trial, Haskell testified that the presence of hovering flies was exactly what one would expect in the first minutes to hours after death, given the ideal weather. "It starts with a few flies," he explained. "Then, those first flies lay down pheromones, a scent that sets up conditions for what we call an oviposition frenzy." Over the next twelve hours, Haskell testified, flies would have literally covered the corpse, pasting the surface of the severed neck with a crust of their creamy white eggs. If the murder had occurred the afternoon before the body was found, those eggs would already be hatching by the time the body arrived at the morgue.

On cross-examination, the DA could only challenge Haskell's experience with "Florida flies." Had he ever conducted field research in the Everglades?

He had not, the entomologist admitted.

"So you can't say with certainty, Dr. Haskell, that there are not areas of the Everglades that might be free of flies?"

"Highly unlikely," Haskell replied. "There are blow flies virtually everywhere on the face of the Earth. A few years ago, I collected them at 18,000 feet in the Himalayas."

A court delay gave Haskell the opportunity he needed to erase any remaining doubt. On November 13, 1992, on the eleventh anniversary of the murder, Haskell placed two freshly slaughtered pigs across the dirt road from the death scene. Within seconds of the pigs hitting the ground, they drew the first flies. By 10 A.M.

the next morning, Haskell had his "oviposition frenzy." Flies and eggs covered both carcasses, with hundreds of tiny maggots hatching by midafternoon, when the medical examiner would have been conducting the victim's autopsy. When the prosecution learned of Haskell's findings, it dropped all charges.

After two decades in business, North America's forensic entomologists all had such tales to tell—stories of pinpoint accuracy borne out by experiment, confession, or other evidence. Most times, the precision resulted from some rare confluence of events such as murder coinciding with a brief window of warm weather in an otherwise cold, and so, insect-free span of time. But such tantalizing success had a dark side, namely the temptation to claim similar precision in other situations.

The temptation to exactness only increased with the development of more sophisticated computer programs for modeling insect development. The most powerful of these factored the predicted arrival and departure times of dozens of carrion insects, from blow flies to hide beetles through mites, against each species' developmental times under varying temperatures and other environmental conditions. If Greenberg's accumulated-degree-hour-approach invited unwarranted precision, computer modeling brazenly seduced it.

Still, as a whole, forensic entomologists resisted the temptation. On the witness stand, they dutifully fuzzed out their results with generous if subjective fudge factors: plus or minus twelve hours, a day, or two weeks—whatever the particular situation seemed to demand. Most were likewise scrupulous about phrasing their time estimates as minimums, as in "the insects show the victim had been dead for *at least* five days prior to its discovery," because unseen barriers and other factors could always delay the start of the insect clock.

The conservatism proved all the more appropriate given that, even at the turn of the twenty-first century, every American forensic entomologist calculated time since death slightly differently.

Some estimated insect development via degree hours. Others preferred the more cautious "degree days." Still others simply rounded up or down from the developmental times of laboratory-reared flies, either their own or those of published authors. A few simply compared the size of the maggots found on a corpse to that of their own reference collection gleaned from their research with pig carcasses or human cadavers.

Unavoidably, the result was the occasional courtroom showdown, with each entomologist furiously punching a calculator to come up with a different time estimate. Given the Dirty Dozen's intimate familiarity with each other's methods, as well as some of their early foibles, their courtroom encounters occasionally degenerated into personal attacks (delivered by feeding professionally embarrassing questions to cross-examining attorneys). For the most part, however, the entomologists and their slightly different methods concurred so long as no one tried to get *too* precise.

Yet if the insect stopwatch had reached the limits of its accuracy, how was the next generation of forensic entomologists to advance their science? Perhaps it was in seeing entomology's place in the larger picture, in the overall *ecology* of the human corpse and the confluence of life that rises from it. In doing so, entomologists would come to marry their forces with a cadre of crime fighters even more unlikely than themselves.

9 Plants, Pollen, and Perpetrators

Pile the bodies high at Austerlitz and Waterloo.
Shovel them under and let me work—
I am the grass; I cover all.
—CARL SANDBURG, "GRASS"

UNIVERSITY OF FLORIDA botanist David Hall had been planning to wind up his informal talk to the state medical examiners' commission with a question-and-answer period. But there was something about the deputy sheriff perched on the edge of the stage that told Hall he'd better cut it short.

It wasn't that Hall was uncomfortable around police types. There was even something mildly *Miami Vice* about Hall's usual uniform of flowered "Don Johnson" shirts and blue jeans, something off-duty cop about his closely cropped crew cut and beefy frame. But there was an edginess about this officer that rattled the botanist. He had first noticed the man jump from his seat midway through the presentation. After slipping out the back door of the conference room, he had quickly returned to sit within a foot of Hall's lectern. Aside from the furious twitching of one foot, he had not moved since or shifted his insistent gaze from Hall's face.

Hall, the director of plant information services at the University of Florida Herbarium, wanted to think he was holding the officer's attention with his handful of forensic yarns. Invited to address the medical examiners' 1982 midsummer conference in Jacksonville, he had led off with one of his best cases, a rape and murder that had made local headlines a decade earlier. The Fort Lauderdale detective overseeing this case had sent Hall an envelope of bark chips vacuumed from the rug and windowsill of the victim's upstairs bedroom. For all practical purposes, it had been no different from the other 5,000-odd identifications Hall made annually. From the diagnosis of oak gall and leaf blight to the classification of Florida's endless parade of plant invaders, the work kept Hall buried under a deluge of mail that arrived by the yard.

The drill in this case had been the same. Hall had lifted the magnifying loupe that perennially hung from the lanyard around his neck and jammed the lens into his eye socket. Then, holding the chunk of smooth, lichen-stained bark a quarter inch away, he focused for one, two, three seconds: "Gumbo-limbo, *Bursera simaruba*."

The Fort Lauderdale investigator was ecstatic, for he had recognized the tall, twisting tree, a native of the Florida coast, growing outside the victim's upstairs bedroom window. Hall's identification confirmed the detective's hunch that the killer had entered the house by way of that tree, for there had been no signs of forced entry, only a partial set of fingerprints—pointing in—on the bedroom windowsill.

A few months later, local police matched the fingerprints to a recently arrested burglar. On questioning, the suspect admitted to having been in the woman's house, even to having had sex with her. He claimed, however, that it had been consensual, no harm done. "We hit it off. She invited me in," he swore. But the bark chips found in the man's pants' cuffs argued otherwise. "Again, gumbo-limbo," Hall testified at trial. The man had not come in through the front door as he claimed.

In the years that followed, the UF botanist cemented his reputation among local law enforcement by fingering a rapist with flower petals on a blanket and a killer with grass fragments wiped from the butt of a gun. Hall related these stories to the medical examiners gathered at the Jacksonville conference center. Yet despite all the true-crime drama, it was during the driest stretch of Hall's lecture that the deputy sheriff had jumped out of his seat and charged out of the room. Hall had been explaining how, in theory, one could use the layers of leaves at an outdoor crime scene to count back seasons since a body or other piece of evidence hit the ground. "Each year's layer can be a different color, or a different species, and sometimes you can just lift those layers up, like pages in a book." Hall knew of forensic pathology's longstanding problems with determining time since death from medical signs. He was trying to make the point that plants, like anything that changes predictably over time, could be used as an alternative clock.

Now, with five minutes left to his allotted time, Hall was pleased to see the flush of hands that shot in the air as he concluded his prepared remarks. But before Hall could answer the first question, the half-cocked deputy was on stage, grabbing him by the arm. "We gotta go, Doc. We may already be too late." Hustling the botanist out of the Jacksonville conference center and into his squad car, the young officer radioed ahead his destination, Gold Head Branch State Park, fifty miles southwest across Clay County.

Hall knew the place, some 2,000 acres of dry, rolling sand hills on Florida's central ridge. A deep ravine divided the reserve where the spring-fed Gold Head Creek carved a meandering path to Lake Johnson. In that ravine, the young officer was now explaining, a crew from the county sheriff's office and the state crime lab were removing a skeleton found by hikers the previous day. "We're damn proud of our record of no unsolved homicides," he explained. "I'm just hoping there's some plant evidence left down there to help us keep it that way."

Forty minutes later, as the botanist and his escort ducked under a ribbon of yellow police tape, they could see that they were indeed too late. The crime technicians were hefting the last of their boxes out of the ravine. All that remained of the body was a greasy partial outline in the underbrush. At least a dozen pairs of shoes had already trampled the area, mixing and obliterating any botanical evidence. Hall had no way to tell which leaves, if any, had lain atop the remains or what new growth might have worked its way through and around the skeleton before it was ripped out of place.

The chagrined crime techs offered photographs and sketches made before they began processing the scene. It wasn't enough. "I'd have to see something in place," Hall explained. Had they already gathered everything? A jaw and a leg bone had yet to be located. "That might be enough," Hall suggested. They called in the National Guard.

With the first light of morning, more than a dozen guardsmen and -women walked through the entire ravine at arm's-length. Some fifty yards from the crime scene, they found the jawbone, and a little farther along, the long shank of a shinbone. This time, the investigator in charge ensured that the evidence remained in situ, untouched. He placed a cardboard box over each bone, posted guards, and dispatched a squad car to the Gainesville herbarium to fetch Hall.

The jaw lay on a bare patch of sand. Nothing he could do with that, Hall realized. The tibia, by contrast, drew a smile. Its long shaft had dropped across a five-inch-tall turkey oak seedling, *Quercus laevis,* bending it to the ground. Removing the bone, Hall saw a fully formed leaf with a distinct black band where the bone had shaded and killed the underlying chlorophyll. This could prove to be informative.

Hall reasoned that the bone had been in place for at least two weeks. Any less time and at least some of the underlying chlorophyll would have remained alive—paler than normal, perhaps, but not dead black. Therefore, the bone had dropped on the turkey oak on or before the last week in June.

How much earlier could it have been? If the bone had dropped before the leaf had completed its development in early spring, the disturbance to leaf growth would have produced a malformation of the growing tip. There was none. So the bone had reached this final resting place *after* the local turkey oaks had finished leafing out for the year.

Hall spent the rest of the morning talking to park rangers and maintenance workers, asking them to think back to the year's first flush of treetop green. Spring usually hit central Florida in mid- to late February, but that winter had been especially cold and long. The turkey oaks, like most everything, had leafed out a bit late, in the first or second week in March. Hall added another three weeks for the leaves to mature, and two more weeks for the bone to kill the band of chlorophyll by depriving it of light. The leg bone had come to rest atop the turkey oak, at the earliest, in mid- to late April.

The next factor in extrapolating time since death centered on how the bone got there in the first place. Hall could see gnaw marks, no doubt from some kind of scavenger. But were they from a bear, which would have been powerful enough to tear into fresh human remains, or a smaller animal that would have had to wait until the skeleton had begun to disarticulate?

A wildlife specialist at the Florida Museum of Natural History identified the tooth marks as those of a large dog, perhaps a German shepherd. After talking with a county medical examiner and reviewing weather data for the previous year, Hall estimated it would have taken the body four to six months to deteriorate to the point where the dog could have yanked free what was left of the leg. The longer end of the spectrum sounded right, given the prolonged winter weather. That pushed the most likely time of death to the end of September.

Hall's estimate proved enough to identify the victim from missing-person reports. But it would be another twenty years before the Clay County Sheriff's Department regained its perfect record for unsolved homicides and closed the case of the Gold Head Branch murder. A guilt-ridden witness finally stepped forward to describe the shooting and name the killer. The victim had been shot, execution-style, in early October. Hall's time-of-death estimate—his first of many, as it turned out—also proved to be one of his most accurate. Extrapolating back over nearly a full year, he had come within a week or two of the actual murder.

At the time, Hall would have never thought to bill himself as a forensic botanist. His police work made up a sliver of his identifications and consultations at the UF Herbarium. Even when Hall struck out on his own to become one of the country's most unusual private eyes in 1991, the bulk of his consulting work involved civil cases such as the identification of endangered species in areas slated for commercial development, or the determination of fault in cases of environmental damage. Yet the FBI already counted him among the small handful of botanists they could call on when some shred of plant material ended up lodged in one of their murder cases.

Hall was not entirely alone in his specialty, although North America's crime-fighting botanists didn't add up to so much as a Dirty Dozen. Most of them had been pulled into the business by university colleagues more directly tied to homicide—professors of pathology, anthropology, and entomology, whose work had trained them to look for the nontraditional clue, and the nontraditional expert, to make sense of it. The work came in spurts too unpredictable to warrant a full-scale field of study. Yet by sharing cases and comparing notes on successes and failures, this handful of crime-fighting botanists created a recognizable field. Like the other "forensic biologies" that came before it, forensic botany had a more interesting history than its practitioners may have realized.

* * *

NO DOUBT, HUMANS have been deducing evidence from plants as long as they've been tracking enemies and animal prey. The first mention of plants in a legal context appears in Plato's *Phaedo,* 399 B.C., which describes Socrates' self-administered death sentence of poison hemlock, *Conium maculatium,* with a graphic account of its associated symptoms of icy hands, muscle paralysis, and coma. Plants would continue to weave their way through legal history in their time-honored roles of poisons and illegal intoxicants. But it was not until 1935 that a botanist took the witness stand to play the evidence-matching game that would make forensic botany famous.

Two years earlier, the infant son of famed aviator Charles Lindbergh and author Anne Morrow had disappeared from his crib. Despite the payment of $50,000 in ransom money, the baby turned up dead, buried in the woods some four miles from their Hopewell, New Jersey, home. The first break in the investigation came when some of the marked ransom money reappeared in circulation. Police traced it to Bruno Hauptmann, an escaped German convict who had slipped into the United States via Canada. Though Hauptmann named another man as the source of the bills and denied any connection with the kidnapping and murder, he would be convicted largely on the testimony of a wood anatomist.

Arthur Koehler, the author of thirty-two scientific articles on the attributes of wood, had offered his services to Lindbergh soon after the kidnapping in 1932. If given a wood sample from the handmade ladder used to reach the baby's nursery, Koehler said, he might be able to trace it to a specific lumber mill. As Julie Johnson-McGrath wrote in the *Legal Studies Forum* (1998):

> Two of the ladder's rails, made of yellow pine, showed marks indicative of the use of a defective planer, and Koehler felt that this was a clue that could be pursued, despite the almost ubiquitous use of yellow pine as a construction material. After months of traveling and correspondence, Koehler located the mill with the defective planer in South Carolina and obtained a list of the lumber yards that had received any of the 47-train car loads of yellow pine it had shipped between 1929 and 1932. Over the next year and a half, Koehler and State Trooper Lewis Bornmann visited the thirty lumberyards in the Bronx, Brooklyn, and other parts of New York and New Jersey that were on the list. After his arrest for passing the ransom money, Hauptmann was identified as a customer of one of these yards. Koehler clinched the case for the prosecution by testifying that the used piece of wood in the ladder came from Hauptmann's attic.

So thorough and credible was Koehler on the witness stand, and so far-reaching the media coverage, that detectives the world over began looking at the green backdrop of their work in a new way. For the most part, cops had a Koehler-style matching game in mind when they showed up at herbaria and botany labs with their envelopes of splinters, seeds, and cockleburs. Could the botanist link some scrap of vegetation plucked from a sweater, shoe, or car grille to something growing at the scene of a crime? Few police realized that the plants insinuating themselves through a murder scene could also freeze-frame time of death, very often to a year or a season, and occasionally to a month, even to a week or less.

Not that it always takes a scientist to read the pages of the botanical calendar. Among forensic anthropologist Clyde Snow's favorite stories was a case involving his identification of human remains found lying in a Louisiana soybean field in the late 1960s. From two burr holes in the skull, Snow surmised that the dead man had wandered away from a local mental institution (he had undergone a prefrontal lobotomy). But the administrators of the psychiatric hospital had the dubious habit of recording unexplained disappearances as routine "discharges" to avoid scandal. With the dry bones revealing little as to time of death, Snow needed some way to narrow his search through the discharge records of the hundreds of patients that had passed through the hospital's revolving door in recent years. Luckily, at one point, the farmer on whose soybeans the patient had landed walked back to watch the hubbub of police activity. He took one look at the patch of dead plants beneath the bones, scratched his head, and said, "By the size of them beans, it hadda been late June or July." It was enough for Snow to identify his man—a patient who had talked incessantly about heading south to track down his mother and "pay her back" for the lobotomy that had put a stop to his teenage rebellion some twenty years before.

The farmer's familiarity with crops, like a park ranger's awareness of trees leafing out, was the sort of intimacy with the land on which early forensic botanists relied in their estimates of time since death. "I used to get my best information by simply walking down a country road and talking to people in their yards," says Hall. Did they notice, for example, when the pines shed their pollen that spring, when the sunflowers reached full size, or when the dandelions went to seed? A prudent botanist has too great a respect for the seasonal and geographic variability of the plant world to presume to know such things from books or charts. Unfortunately, by the mid-1980s, urban sprawl was rapidly robbing forensic's newest field of a vital resource. City dwellers and suburbanites simply were not as in tune with their surroundings as their rural predecessors.

So it was with great respect for the limitations of their science that in the last decades of the twentieth century a handful of American botanists offered their services in the age-old struggle to pinpoint time of death. The botanical snapshot of Hall's bone on the turkey oak and Snow's body on the soybeans represented just two of the ways that their expertise could be harnessed to the effort. The greatest advantage of this method of freeze-framing death's season lay in its simplicity. Any Sherlock Holmes fan could appreciate the cleverness of dating a deathbed of linden blossoms to June, just as the jury in a Colorado murder readily accepted testimony that frostbitten leaves spilled into a hastily dug grave pinned murder to an October cold snap. In large part, it was this commonsensical aspect of plant evidence that greased forensic botany's acceptance in the courtroom. It proved a great advantage in circumventing the gauntlet of legal challenges that opposing attorneys could throw down to block scientific testimony not to their liking. This isn't to say, however, that forensic botanists didn't break new ground, or that they didn't have to patiently walk judge and jury through some of their more complicated methods.

Such a case unfolded before a world audience in 1997, when plant evidence helped establish for the history books the perpetrators of an atrocity unearthed in eastern Germany three years earlier. Construction workers had discovered an ungodly pile of bones while digging the foundation for a new building in Magdeburg. Local speculation quickly ran to that of a mass suicide involving none other than Adolf Hitler, rumored to have slipped into the city in the chaotic months following Germany's World War II surrender.

Forensic anthropologists ruled out that possibility. They tallied the remains of thirty-two bullet-ridden young men, with bone development indicative of late teens to mid-twenties. The bones' *postmortem* age proved more elusive. Dry, but not yet crumbling to the touch, their texture suggested that the victims had died sometime between 1945 and 1960. This left open the possibility of mass murder at the hands of one of two infamous organizations that had once controlled the city.

The Gestapo ruled Magdeburg from the late 1930s through May 1945, when they beat a chaotic retreat in front of advancing Russian troops. Before doing so, they might have executed any remaining prisoners. By the 1950s, the Gestapo's thuggish grip on Magdeburg had been replaced by that of SMERSH, the Soviet counterintelligence agency that chose the city as its headquarters. (The acronym derives from the Russian phrase "death to spies.") The town's young people had chafed under SMERSH's repression and arbitrary brutality, and in late June 1953, they rebelled. Amazingly, a platoon of Soviet soldiers refused orders to fire on the crowd. It was the last anyone saw of the young troopers.

German prisoners in the spring of 1945, or Soviet soldiers in the summer of 1953? For answers, officials turned to Reinhard Szibor at Magdeburg's newly opened Otto von Guericke University. In the German tradition of *forensic biology,* Szibor's expertise spanned both the plant and animal world. In particular, he was famous for using pollen to link criminals to their crimes. As any hay fever sufferer can attest, the male gametophyte of flowering plants saturates virtually every corner of our world with its light, windborne grains. Nothing escapes this pollen rain, a fact that archaeologists and geologists have long used to reconstruct the vegetation of prehistoric eras. In his earlier, more traditional forensic work, Szibor had used pollen to place a suspect at the scene of a crime, by matching grains found on the suspect's hair, skin, or clothing to the vegetation at the scene.

In theory, Szibor reasoned, he should likewise be able to use pollen to determine the "when" of a crime, in particular whether the mass murder took place when Magdeburg's trees were in full spring bloom or when its fields lay under a carpet of summer weeds and grass. "When any of us die, we do so with pollen in our noses," he explained. Virtually indestructible, that pollen would remain long after the nasal membranes had turned to dust.

To test the idea, one of Szibor's graduate students tracked the pollen in his own nose over the course of a year. His "handkerchief test" consisted of weekly blowings, followed by pollen identification with light and scanning electron microscopes. True to expectations, the pollen in the student's nose precisely tracked the blossoming of local alder, hazelnut, willow, and juniper in spring, then switched to that of summertime plantain, sagebrush, lime tree, and rye.

Their theory confirmed, Szibor's group returned to the victims' remains and carefully rinsed out the nasal cavities of twenty-one skulls. Seven yielded significant amounts of plantain pollen, with smaller quantities of rye and lime. To verify that the microscopic grains had not simply settled on the victims as they moldered in the grave, Szibor performed similar tests on a sampling of their other bones. None yielded a significant amount.

The pollen in the nasal cavities pointed decisively to a SMERSH atrocity. Corroborating the finding, Szibor noted signs of extensive dental decay but little dental work on the victims' teeth—all too typical of Soviet soldiers during the postwar years.

* * *

IN ALL THE above examples, botanists used static evidence such as pollen grains or dead plants to provide a botanical snapshot of month or season of death. In theory, plant growth *since* death could give them an even more direct yardstick of postmortem interval, albeit one that demanded great expertise to measure. Clearly, a plant going about its sunlight-driven work of transforming carbon dioxide to carbohydrate is a living clock of the truest kind.

Indeed, the original concept of "accumulated degrees" used by entomologists to measure insect growth over time came from early-twentieth-century agronomists who needed a way to calculate when to plant a crop so that harvest fell during optimal weather. What they discovered was that plants, like insects, depend in large part on heat to drive their growth. Consequently, today's farmers have agronomy almanacs filled with tables specifying the total number of degree days required for a given crop to mature from seed or blossom to harvest. Unfortunately, countless other variables such as sun exposure, soil quality, precipitation, and insect attack can skew plant growth beyond the parameters set by simple mathematical formulas.

Fortunately for forensic botany, certain plants come equipped with their own internal chronometer, a woody calendar of sorts. As any child fresh from summer camp can explain, a tree stump displays its age in the concentric rings radiating out from its core. Woody plants normally begin to produce these rings in their second year of growth, when the cells in a cylinder of plant tissue called the *cambium* begin to divide. Typically, spring's first flush of wetness and warmth wakes the cambium from its winter slumber. It enters a period of furious cell division. With most cell divisions, the outer of the two daughter cells takes over its mother's role as cambium, while the inner cell adds itself to the mass of wood, or *xylem,* lying just inside. By nature's design, the xylem cells quickly hollow out and die, leaving behind their tubelike skeletons to serve as the tree's plumbing, conducting moisture skyward from the roots through capillary action.

But not all xylem cells look alike. The early frenzy of spring growth produces especially large, thin-walled cells. The result is a pale band of tissue called *earlywood,* or *springwood.* As summer progresses and moisture and daylight wane, cell division slows to produce progressively smaller and thicker-walled xylem cells, until finally growth stops in the fall with a line of crumpled dead cells. When growth kicks off again the next spring, the contrast between fall's dark *latewood* and the new season's earlywood can be seen as a stark line of demarcation. All this was understood by botanists of the nineteenth century. However, it was not until 1928 that an archaeologist realized how to read past history in the annual rings of a tree.

American archaeologist Andrew Douglass noted that trees tend to produce especially wide annual rings during especially wet seasons and narrow ones during drought. He likewise noted that spring cold snaps, Indian summers, and other fickle weather could produce extra, or "false," growth rings throughout the year. Fires, insect infestations, and other disturbances could likewise leave their mark on the woody record. Finally, Douglass realized that he could correlate a tree's ring patterns to specific years by studying a region's weather records as well as historical accounts of fires, pestilence, and the like.

In 1929, Douglass applied his new technique of *dendrochronology* to timbers pulled from the prehistoric Pueblo ruins of New Mexico's Chaco Canyon. He described his work in an article he wrote for *National Geographic:*

> By translating the story told by tree rings, we have pushed back the horizons of history in the United States for nearly eight centuries before Columbus reached the shores of the New World, and we have established in our Southwest a chronology for that period more accurate than if human hands had written down the major events as they occurred.

By the end of the twentieth century, dendrochronology had become a worldwide science with over a thousand practitioners trained to read the stories that reside in living trees and ancient timbers. Most were archaeologists like Douglass, more interested in prehistoric stories of climate change and shifting civilizations than the sordid details of murder. Their techniques did not always translate to the more recent past. Growth changes caused by random events could be especially difficult to disentangle from seasonal rings in the fast-growing and short-lived brushwood typical of civilization's fringes—the kind of vegetation where the missing dead tended to turn up. Sometimes, however, death leaves its own distinctive mark in the dendrochronological record.

In 1985, such a case ended up in the hands of Regis Miller, a wood anatomist plying his trade in the same USDA Forestry Service lab as the legendary Koehler, of the Lindbergh baby murder trial. In April of that year, Miller received a call from the Alabama State Crime Lab in Mobile. Hunters had found a skeleton at the foot of a tall, spindly tree in the Appalachian foothills northeast of Birmingham. The state police suspected a hanging, because nearly twenty feet above the ground, they could see a length of shirtsleeve knotted around the slender trunk. Yet they remained clueless as to whether the man had committed suicide or had been lynched. For that matter, they could not figure out how or why he came to be strung from so high a perch. Could the noose have been growing skyward in the years since the hanging? Could Miller tell by its height how long ago that might have been?

First, Miller set straight their misconception that something tied to a tree would grow upward. Trees, like all plants, lengthen only from their tips after the first year's growth, which is why a sign nailed into a tree at eye level stays there year after year. Still, the investigator hung on the line, hoping for some hint to help him narrow his search through missing person reports. Forensic anthropologists could tell him only that the man had been dead somewhere between one and fifteen years.

"Well, send me as much of the tree as you can," said Miller. "At the least I can identify it for you." In the days that followed, Miller and his colleagues debated how the jury-rigged hangman's noose could have ended up twenty feet in the air. Miller recalled how as a boy, he and several of his buddies would shinny up a medium-size sapling to bend it to the ground, then take turns climbing on board for a catapult ride when the others let go. "Maybe that's how this guy hung himself," he proposed.

Certainly there would have been easier ways for a lynch mob to do the job. However, someone desperate and alone might have gotten inventive. Judging from the width of the trunk shipped to them, Miller figured that the tree would have been spindly enough, a few years back, to bend under the weight of a full-grown man. "So he could have climbed up, tied one end of the sleeve around his neck and the other around the trunk, and then released his grip." Perhaps the tree sprang back, breaking his neck. Perhaps he just dangled there.

When the length of tree trunk arrived at the lab, Miller immediately recognized it as cucumber tree, *Magnolia acuminata,* a fast-growing Appalachian magnolia named for its green, oblong pods. However, what really grabbed his attention was the shirtsleeve noose, which the crime techs had left knotted in place where it girdled the trunk. The surrounding bark had partially overgrown it, and Miller could see an obvious indentation in the girth of the trunk. He reasoned that if the fabric had been tied snugly enough to suspend a man from the neck, it would have taken at most a year, probably less, for the loop to begin to constrict the trunk's expansion.

With a small power saw, he cut several disks through the trunk at, above, and below the sleeve. Not surprisingly, the constriction had mashed the outer growth rings together like a squashed layer cake. Miller looked at the cross sections first with a hand lens and then stereo and light microscopes. Hours later, his eyes burning and temples pounding, he walked down the hall to get his colleagues' opinions. The consensus: Five tight but discernible lines extended inward from the dented bark and cambium before the growth rings opened up to form normally spaced annual bands. The man had died five years earlier.

Miller had yet to hear whether his time of death estimate had proved useful in solving the mysterious hanging when an even more quirky slice of tree arrived in his lab the next year. Utah police wanted to know if the wood anatomist could give them a kind of "time of death once removed." Hikers in the mountains east of Bountiful, in northeast Utah, had led police to an aspen tree bearing the carved inscription "Ted Bundy 78." Perhaps it was the handiwork of someone unlucky enough to share the name of the most frightening serial killer of modern times. Or perhaps it was a sick joke.

But no one in the Utah State Crime Lab was laughing. Bundy had torn a murderous path across the state in the fall of 1974.

Twelve years later, anxious parents still checked on their sleeping daughters in the night.

The "Bundy Tree," as it had quickly become known in the local media, raised especially painful memories of a pretty teenager whose body police had never found. Debbie Kent disappeared from the parking lot of Bountiful's Viewmont High on a November evening when a stranger later identified as Bundy was seen lurking around the school. Police suspected that the serial killer dumped Kent's remains in a national recreation area east of town, precisely the same area where the carved aspen now stood. The public was demanding that police launch another, more massive search of the surrounding wilderness for Kent's body and any other victims.

However, if the number cut into the tree in fact signified the date of the carving, it could not have been the work of *that* Bundy. On February 15, 1978, police arrested the serial killer in Florida, where he had been preying on girls and college coeds since his prison break the previous summer. The trouble was, there existed another, far grislier explanation for the number. When he was on death row, Bundy abandoned his earlier claims of innocence and began confessing to more and more of his killings. By the time of his execution in 1989, his personal tally had reached twenty-eight. Police counted at least thirty-six murders and suspected far more. Though it boggled the mind to consider the possibility, some were saying that the "78" carved into the Utah aspen might represent Debbie Kent's place in Bundy's actual body count. The state crime lab was sending Miller the section of tree trunk bearing the disturbing carving with the hope he could resolve the speculation by determining when it had in fact been carved in the tree.

Miller began, as before, by cutting cross sections through the wood at, above, and below the telltale damage. He immediately saw that the vandal's knife had failed to penetrate to the depth of the tree's woody xylem rings, or even nick the vascular cambium that produced them. As a result, he lacked any clear marker to use as a reference point to count back through the annual bands of growth. All he had was a superficial bark wound.

Then again, Miller worked in one of the few laboratories in the world with its own bark anatomist. Under the guidance of Ray Evert, Miller examined the damaged bark more closely. Both men understood that as a young tree expands in width, its growth eventually stretches its epidermis, or "skin," to the bursting point. The resulting exposure to drying air and insects would quickly kill the plant if not for its quick adaptive response. A single layer of cells just under the broken surface begins to divide. Botanists call this thin cylinder of growth the *periderm.* With each division, the outer of the periderm cell's two daughters becomes a cork cell—thick-walled and impregnated with wax to guard against water loss and insect attack. As the tree trunk continues to expand, these outer cork layers likewise rupture, triggering the formation of newer and younger cork layers beneath them. The end product is bark, its roughened texture the result of year after year of ruptured cork layers and the formation of new cork within.

The question in Miller's mind was whether the bark damage produced by the vandal's knife might trigger a similar protective response, a so-called wound periderm, which might then begin forming successive annual layers. He and Evert examined the normal areas of bark, then the bark directly beneath the carving, and finally the areas in proximity to it. Excruciating barely describes their eyestraining work. Still, under just the right light and focus, the lines of periderm snapped into view under the lens of the laboratory's best light microscope. Counting the lines of periderm away from the scar and those nearest to it, the botanists counted a maximum difference of eight microscopic layers. That is, eight wound periderms. The number gouged into the tree did in fact represent an accurate date, they concluded, carved in the summer following the serial killer's final arrest in Florida. The Utah State police called off their search of the surrounding wilderness, and the state's citizens returned to the work of healing less tangible scars.

* * *

UNDER THE RIGHT circumstances, a dying plant can tick off time since murder as well as if not better than a growing plant. When Ralph Takemire decided to dump his girlfriend's body in a ditch alongside a Colorado sunflower field in July 1991, he provided police with just such a postmortem clock. Takemire covered the corpse with an armload of uprooted sunflowers. The camouflage worked well enough. Terra Ikard's body remained unnoticed by passing traffic until a highway maintenance worker chugged up on his tractor-mower for the monthly clearing of the embankment. As any county sheriff will tell you, the men and women who mow America's roadsides each summer deserve to be deputized for all the corpses they uncover each year (hunters and backpackers do almost as well).

The second Arapaho County employee on the scene that day was crime tech Jack Swanburg. Climbing out of his evidence van, Swanburg immediately recognized the sickly sweet smell that told him the medical examiner would be of little help determining time of death. Beneath the blanket of wilted, four-foot-tall flowers, he saw the bloated form of a woman in the early stages of putrid decay. The sunflowers had fared better, something your average crime tech would not have noticed.

But Swanburg had been spending his weekends with a thought-provoking group, a quirky collection of University of Colorado scientists and law enforcement people who spent their time planting pigs in a field outside Denver. In the tradition begun by forensic entomologists nearly a decade before, the group had adopted freshly slaughtered pigs as their homicide stand-ins. But rather than leave their test subjects to the open air, they buried them . . . then watched for the signs that could betray the most careful killer. A subtle slump in the soil as the decomposing chest cavity sags and collapses beneath the sod's weight. The claw marks of scavengers sensitive to the tantalizing scent of what lies below. A green flush of telltale growth as weeds spring up from the disturbed soil. NecroSearch, they called their group when advertising their free services and morbid passion—finding the missing and presumed dead.

In addition to Swanburg, NecroSearch's founding members included Jane Bock, an elfin plant ecologist at the University of Colorado. She was nearing retirement when a pathology professor at the university's medical school first introduced her to what would become an unexpected sideline. Naively, Bock had believed the pathologist when he promised he would never send her anything more ghoulish than a slide of plant cells . . . albeit plant cells taken from a murder victim's stomach. Still, the stories of lives cut short had a way of pulling Bock in, until she found herself spending more time in Swanburg's pig patch than in the neglected flowerbeds in her own backyard.

It was Bock whom Jack Swanburg called as he walked away from the roadside ditch that summer morning. Back at his evidence van, he punched in the number to her office on the chunky cell phone mounted to the dash. "I've got a big bouquet of flowers for you," he said, before explaining what he had found covering Colorado's latest murder victim. For comparison, Bock asked him to pick a second bunch. "Write down what time you pull them out of the ground," she instructed.

Processing the crime scene took several hours, including the body's delivery to the morgue. By the time Swanburg merged into the westbound lane of US-36, he could see the sun sinking low over the red rock cliffs that marked his destination thirty miles to the west, the verdant campus of the University of Colorado at Boulder, nestled against the Front Range of the Colorado Rockies.

Bock was waiting, happy to see her beefy compatriot in arms until she got a whiff of the burial bouquet in his jumbo-size evidence bag. The rank smell resembled nothing even remotely botanical, nothing Bock had ever smelled in her life. Still, she did not have to ask, nor did she want to discuss it. Quickly resealing the bag, she stuffed the rank-smelling sunflowers into the biology department freezer, then carried the less-fragrant and more recently picked flowers to her rooftop greenhouse. Over the next hours and days, Bock explained, she would watch these still-fresh sunflowers wilt. When they matched the killer's bouquet in her freezer, she would have her estimated time since death.

As luck would have it, the climate-control system of the biology department's rooftop arboretum had broken down the week before. So Bock's control flowers would experience the same natural fluctuation of daytime heat and nighttime chill as did the sunflowers used to cover the body.

Bock began a daily ritual in which she would open her freezer each morning to refresh her memory of the burial bouquet's appearance, then climb the stairs to the roof to check the controls gradually wilting in the summer heat. After seven days, the leaves of the greenhouse flowers took on the same limp appearance as those in the freezer. They remained in that droopy but still supple state for another six days, before suddenly shriveling.

When Bock telephoned Swanburg with her one- to two-week time estimate, she learned that her morbid nosegay had an equally repulsive counterpart. After dropping off the sunflowers in Boulder, Swanburg had swung northeast to Colorado State University in Fort Collins, where he had left a small carton of maggots with entomology professor Boris Kondratieff. Kondratieff's estimated postmortem interval matched Bock's exactly.

In the months that followed, their determinations would play a significant role in the arrest and conviction of the victim's boyfriend, who had apparently killed her as part of a plan to sell their baby to a couple in Kansas. More important, said Swanburg, "while most of us in law enforcement were familiar with the bug stuff, no one that I knew, at least, had ever used the condition of plants to determine PMI (postmortem interval). It became a proving case that put Dr. Bock's name in the minds of law enforcement and pathologists across the state."

Yet Bock had already proven key in a handful of cases involving a very different kind of postmortem marker, one that would go far to rehabilitate what had become one of forensic science's most controversial and discredited methods of determining time since death.

10 The Pathologist's Garden

Miserable mortals who, like leaves, at one moment flame with life, eating the produce of the land, and at the next moment weakly perish.

—HOMER, CIRCA 700 B.C.

MEDICAL EXAMINER BEN Galloway had always enjoyed the contrast between his intensely focused work in the steel-and-tile autopsy suite of the Jefferson County Coroner's Office and that of his freewheeling banter with students at the University of Colorado School of Medicine, a few miles away. The latter sparked his imagination, kept him sharp. The former gave him the concrete sense of purpose that came with bringing closure to families torn apart by sudden death under violent or simply unclear circumstances. Not that Galloway could always remain as objective as his personal philosophy implied. There were faces that haunted him.

A few months before she ended up on Galloway's autopsy table in the fall of 1982, the fresh-faced young woman had graduated from a small liberal arts college in Massachusetts. She had come to Denver to intern at a local radio station. According to police trying to piece together the victim's last hours and days, the young woman commuted to work each day on a bus to and from her aunt and uncle's home in the suburbs south of the city. They had been quick to report her missing when she failed to return home one Friday night. Her body turned up before morning, dumped by the side of an unlit road a few miles away.

In the days that followed, police quickly focused their suspicions on the intern's boyfriend, the last person to see her or be seen with her, when they met for lunch at a McDonald's across the street from the radio station where she worked. The mere fact that he was the victim's boyfriend made the young man a prime suspect, for as anyone involved in homicide investigation can attest, the most dangerous guy a woman is likely to encounter is the one who says he loves her. But clearly, the DA would need more than that sad fact to build a murder case. Their hopes were with Galloway and his estimated time of death. Death in the afternoon placed the couple together. Anything after nightfall removed the boyfriend from suspicion. He had ample witnesses to attest to his whereabouts from dinnertime through the next morning.

The wide time ranges for the pathologist's postmortem markers of algor, rigor, and livor mortis proved no help. That left the dubious postmortem indicator of stomach contents. Galloway had retrieved over a pint at autopsy. But even the general rule of thumb that the stomach empties within two to four hours after a meal told him nothing unless he could discern just what he saw floating in the murky soup. Was it the hamburger and milk shake her boyfriend said she had for lunch? Several witnesses had corroborated his account. Or was it the remains of dinner, pushing death into the evening hours? Galloway could make out something saladlike in the mix. Then again, it could be the lettuce and onions from the hamburger. He just could not tell.

The case continued to intrude on Galloway's thoughts, even on his days at the medical school. Pushing away his class notes one day, he pulled out the university's thick course catalog and began thumbing through its listings. He didn't know exactly what he was looking for until he found it—a third-year course in plant anatomy taught by botanist Jane Bock of the school's Department of Environmental Biology. Dialing Bock's office, Galloway got right to the point, describing the jar of evidence sitting at the Jefferson County morgue.

"Oh, Lord, no. Don't you send me anything of the kind," the botany professor sputtered. Galloway countered with a quick promise to prepare the microscope slides himself. "All I'll be bringing you is a smear of plant specimen, maybe a half-dozen slides." Then he delivered his clincher. "This girl was fresh out of college, twenty-one years old, with her whole life in front of her." The pause on the other end of the line told Galloway he had found his mark. Bock saw the faces of her own bright, young students flash through her mind. It could have been any one of them, she thought.

"All right. I'll see what I can do."

Bock hung up the phone with a slightly shaky hand, immediately questioning the wisdom of her decision. She felt utterly unqualified. Her specialty was plant population dynamics, for goodness sake. Before the afternoon was out, she was sharing her dilemma with David Norris, the endocrinologist with whom she cotaught a first-year course in general biology. The two had little in common in terms of research interests, yet they worked well together. In fact, rather than dividing their biology course in the usual manner, so many weeks apiece, they enjoyed lecturing jointly. Bock would begin by describing the basics of a particular plant group, and then Norris would pick up by discussing the plants' physiological effects on their "consumers," be they insects, cows, or humans.

Would it be such a stretch, Bock asked, for her colleague to use his physiology background to shed light on how the human digestive system might affect plant material? It didn't hurt that the two professors had long shared a passion for literary murder mysteries. Bock didn't have to ask twice.

The next day found the two CU professors working late into the evening, staring down the double barrel of a stereoscope as they perused the remnants of the victim's last meal. Norris had already confirmed Bock's hunch that the plant material, at least the individual plant cells, would be the one part of the partially digested soup that should emerge from the stomach's acidic soup relatively unaltered.

Although the fragile membranes of an animal cell begin to rip and burst within moments of its death, the rigid scaffolding of a plant cell remains intact even after its living contents have died. Indeed, a plant's cellulose cell wall remains indigestible to all creatures except ruminant grazing animals and a few highly specialized insects such as termites. Archeologists and paleontologists have long used this enduring quality of plant cells to glimpse the diets of ancient peoples and prehistoric animals through their *coprolites,* or fossilized feces. Yet parsing the vegetable content of a human being's last, partially digested meal would prove to be surprisingly difficult.

For starters, Norris and Bock could find little in the scientific literature to help them identify what they saw through their microscopes. Bock's plant anatomy textbooks contained a few pictures of onion cells—translucent little rectangles, lined up end to end and layered as tightly as a sturdy brick wall. These they quickly found in the mix of cells on Galloway's slides. Clearly, onions would be consistent with what one would find on a fast-food hamburger.

But what were the dimpled hexagon-shaped lumps that seemed to rise in 3-D, beneath the low-power lens of their microscope? What were the pinkish masses of sharp-angled cells, each with six, seven, and eight walls to its cellular box? If the color they were seeing was that of a plant pigment, or *carotenoid,* it would suggest that the girl had not been dead long, Bock mused. Norris agreed. The indigestible cellulose wall might protect the plant cell itself from destruction, but it would not stop the stomach's acid bath from bleaching out the delicate chloroplasts that contain the cell's photoactive pigments.

Still, without a reference set for comparison, the professors had no way to put names to the jumble of mystery cells. They would have to make their own. Bock went produce shopping at a local supermarket, and the next day she and Norris took turns chewing up her purchases and smearing the mush across microscope slides.

Even Bock found herself amazed at the variation they saw beneath the low-power lens of their microscope. The specimens differed dramatically. Lettuce, for example, resembled nothing so much as an elaborate jigsaw puzzle with no two pieces exactly alike. By contrast, cabbage came into view all crazy-angled pentagons, hexagons, and heptagons. "You could project these slides on a wall, and a jury could see the difference from across the room," Norris observed.

Their identifications of the stomach contents followed: The pinkish, saladlike material proved to be red cabbage; the dimpled hexagons, kidney beans. In addition, the victim had eaten onions, tomatoes, and green peppers. They had the remains of a veritable salad bar, clearly not the remnants of a burger at McDonald's.

The police immediately dropped their investigation of the boyfriend. A far more disturbing suspect would enter the picture with the later arrest of a Texas man suspected in a string of murders across the Southwest. When shown a photo lineup including the suspected serial killer, waitresses at a local Wendy's reported having seen him eating with a young woman fitting the victim's description on the night of the murder. A survey of the restaurant's salad bar turned up everything in Norris and Bock's checklist.

In the months that followed, Galloway grew bolder with his gifts to the professors. First came the sealed test tube of partially digested plant material, then swatches of clothing and blankets covered with vomit. Eventually, he was sending bottles containing the entire stomach contents removed at autopsy. Bock grew somewhat inured to the specimens. However, both she and Norris realized their new sideline was encroaching on their primary research interests. They resolved to write and illustrate a laboratory manual that might guide pathologists through their own plant cell identifications.

The National Institute of Justice eagerly ponied up the funds for the identification guide, including a small stipend for a graduate student assistant, who became the project's designated chewer. From the masticated plant material, Norris snapped eerily beautiful micrographs for identification keys: the needlelike crystals characteristic of pineapple; the jewel-like clusters of stone cells that identify pear; and the pouting lips of the stomata, or breathing pores, that stud a leaf of spinach. Even more haunting were the scanning electron micrographs provided by CU plant systematist Meredith Lane. Beneath the electronic needle of Lane's SEM, strawberry seeds loomed like moonscape boulders, massive oil droplets crawled amoebalike over the cratered surface of chewed olive, and cherry pulp spilled its jumbled guts over the fruit's unmistakably leathered skin.

The identification guide proved a hit, every copy snapped up soon after its 1988 publication. However, it did not stem the tide of forensic plant material arriving at the CU environmental biology lab. For the most part, Norris and Bock made their identifications, submitted their reports, and never heard what became of them. But occasionally a detective or district attorney would pull them into a real-life murder mystery as twisted as any of their favorite Patricia Cornwell whodunits.

In 1995, Norris was called to testify at the murder trial of a one-time fashion model at the end of a string of at least nine marriages, several of which overlapped with one another, and two of which ended in murder. Jill Lonita Billiott-Ihnen-Moore-Coit-Brodie-Dirosa-Metzger-Steely-Boggs-Carroll somehow escaped suspicion for the shooting of husband No. 3. She was not so lucky when on the afternoon of October 22, 1993, police found husband No. 8, hardware store owner Gerry Boggs, in a bloody heap on the utility room floor of his Steamboat Springs home. Jill had already moved on to husband No. 9 and a subsequent boyfriend. Yet she remained in Steamboat Springs, embroiled in a bitter series of suits and countersuits with Boggs over who owed what to whom following their rocky nine-month marriage, annulled after Jill "Steely" revealed she had never finalized her divorce to husband No. 7.

Autopsy revealed that the victim had been electrically stunned, beaten with a shovel, and shot multiple times in the chest. On the autopsy report, county medical examiners marked time of death "unknown." From the victim's stomach, they removed over a pint of gastric contents, which they described as having the appearance and consistency of "noodles." It was a description that would come back to bite the prosecution.

In many ways they had a solid case against Boggs's ex-wife, including a stun gun found in her red sports car and two people who would testify that she had previously approached them to commit the murder for her. In addition, Jill "Coit," as she had gone back to calling herself, could not corroborate her alibi for the four hours immediately after Gerry Boggs left his hardware store for the last time, just after 1 P.M., October 21, the day before he turned up dead. That's where the noodles in Boggs's stomach became a sticking point.

The defense could produce witnesses to corroborate Coit's claim that she was with her boyfriend that evening, 160 miles away at the Cactus Moon nightclub in Thornton, Colorado. Noodles, the defendant's attorney argued, pointed to Boggs still being alive at dinnertime, putting Coit in the clear. Backing up the argument, everyone who worked at Boggs Hardware could attest to the fact that their boss never ate lunch. A creature of habit, he opened the store at 10 each morning, then walked two doors down to The Shack, the diner where he always ate the same breakfast of eggs, toast, and hashbrowns, day in and day out.

Could the so-called noodles described by the deputy coroner in fact be those morning hashbrowns? the district attorney asked

Norris, who had already received the stomach contents from the coroner's office. Norris assured the prosecutor that the identification should be straightforward.

Smearing a minute amount of the partially digested foodstuff across a glass slide, Norris slipped it under the low-power lens of his microscope and cranked up the piercing light beneath the stage. A jumble of irregular hexagons snapped into view—bent and deformed to fit their space. The mere presence of intact cells told Norris he was not looking at noodles, for the fine grinding of wheat flour destroys all cell structure. The muddle of cells could be potato or apple, which share nearly identical cell structures. Distinguishing the two would prove easy enough. From a nearby rack, Norris grabbed a small dropper bottle of potassium iodine, a standard chemical test for the presence of starch. The instant he applied a drop of the reddish liquid to the specimen, it flushed an inky bluish black—a strong positive reaction that confirmed "potato."

Not yet satisfied, Norris returned to the jar of stomach contents, for he had seen something besides the whitish strings, something smaller and chunkier. He found what he was looking for in the next slide: row upon neatly stacked row of little translucent bricks—onion cells.

"Hash browns with onions," Norris informed the DA, who called back that afternoon.

"Are you sure?"

"Without a doubt," Norris assured him.

The cooks at the Steamboat Springs diner swore they *never* put onions in their hashbrowns, the DA informed him.

Unshaken, Norris asked Bock to verify his identifications. She concurred 100 percent, though she knew they would not win any points with local law enforcement. If the plant material in Gerry Boggs's stomach didn't come from his usual breakfast at The Shack, the defense had a good argument that he'd eaten again elsewhere, pushing time of death into the hours for which Coit had a solid alibi.

The next day Bock returned to her colleague's office with a suggestion. "We need to approach this as scientists," she said. "If this was a controlled experiment, what's the first thing we'd do?" Prepare a control, they agreed.

The next morning Detective Rick Crotz of the Steamboat Springs Police Department made a point of eating breakfast at The Shack. Taking a seat that faced the diner's open grill, he ordered the same breakfast that had been Boggs's 10 A.M. ritual virtually every morning of his adult life. Crotz saw that, true to his word, the cook did not add onions to the handful of hashbrowns he threw on the griddle.

But when the waitress delivered his breakfast, Crotz saw something glistening in the shredded potatoes. A fried onion. Shifting his gaze back to the grill, Crotz saw a flurry of activity that explained everything. Every time the cook turned an omelet, flipped a burger, or scattered an order of corned beef hash, diced onions skittered across the griddle to mix with other orders.

It would be over a year before Norris received the subpoena to present his results in court. Previously, Bock had always been the one to present their findings, given her botanical credentials. However, Norris had performed all of the gastric analyses on Gerry Boggs himself. So through the haze of a miserable late-winter cold in March 1995, Norris drove the winding mountain roads up from Boulder to the Grand County Courthouse in Hot Sulfur Springs. It would prove to be one of the more disturbing days in his forensic career, for next to him in the witness holding room sat both husband No. 7, who considered himself lucky to be alive, and Coit's twenty-nine-year-old son, who would testify against her. Indeed, the shaken young man had concluded that his mother had likewise killed his own father some twenty years before. The jury found Coit guilty on all counts. She remains in federal prison without the possibility of parole.

11 Chemical Clues

Speak to the earth, and it shall teach thee.

—JOB 12:8

ON THE EVENING of October 20, 1992, another road-weary motorist heading east out of Knoxville steered his car onto the dimly lit side road of Cahaba Lane to relieve himself. The odor hit him as soon as he walked behind the highway billboard that rose high above the lane's backdrop of trash-strewn woods. But it wasn't the usual ammonia reek so typical of such spots. It struck him more as sickly sweet. He nearly screamed when he realized what he was close to standing on—half-unearthed by scavengers was an arm, or what was left of it, curving down to a mat of tangled hair and the blackened balloon of a face.

The metro desk at the *Knoxville News-Sentinel* picked up the police dispatch just in time to make the morning edition. With the hills around Knoxville coughing up some forty to fifty bodies a year, the story warranted no more than a brief "police-blotter" report. Nonetheless, it grabbed the attention of one particular segment of the *News-Sentinel*'s diverse readership. Dressed far too scantily for the gusty autumn day, the women huddled close as they passed the news. They knew the place. They knew the victim. And they had a damned good idea of the killer's identity.

Most every streetwalker in Knoxville knew the pudgy-faced john. A few even knew that his name was Thomas Huskey and that he had literally grown up in the Knoxville Zoo, where his father trained elephants and he did odd jobs until the keepers fired him for abusing animals. Most of the ladies simply knew him as the guy who liked to have sex behind the zoo's maintenance grounds. "The Zoo Man," they called him.

He'd seemed harmless enough until recent months, when he began taking his "girls" to a new place, Cahaba Lane. Two women who had been there now realized how lucky they'd been to make it home again. Huskey had struck one so hard she blacked out. When she came to, he was dragging her into the woods, where he brutally raped her before she managed to break free. Huskey tied up the next prostitute he took to Cahaba Lane, beat her, sodomized her, and left her naked and bleeding in the underbrush. She eventually worked her way loose, found some clothes, and limped to the nearest open shop to call the police. The detective had given her his card, which she now fished from her purse.

Police took Thomas Huskey into custody that night, Wednesday, October 21. The next morning, state anthropologist William Bass led a search of the woods behind Cahaba Lane. The body count was up to three when Bass relayed a message to the Knoxville medical examiner to get forensic entomologist Neal Haskell on the next flight out of Rensselaer, Indiana. Meanwhile, Bass had a new trick up his sleeve.

It had been just a month since the *Journal of Forensic Sciences* published the biggest news to come out of Bass's research facility since 1982, when he and Rodriguez first unveiled its grisly purpose. (By the 1990s, the world would know the place as the Body Farm, thanks to the title of a Patricia Cornwell best-seller inspired by Bass and his grisly work.) Meanwhile, a steady stream of scientific papers had marked the progress of the facility's human-decay research. Like anything to do with the tricky art of gauging time of death, the reports caused a stir in forensic circles. Yet those in the business reserved their keenest interest for anything coauthored by Bass. Indeed, there was no better way for a graduate student to make his or her debut in the world of forensic anthropology than to appear alongside that name in the scientific literature. Of the fifty-odd certified members of the American Board of Forensic Anthropology, over half had been trained by Bass. And of the dozen or so seminal papers on human decay, virtually all of them came out of the UT professor's unique al fresco mortuary.

The Body Farm's 1982 offering continued the tradition, though only the steeliest of death investigators could appreciate its title—"Time Since Death Determinations of Human Cadavers Using Soil Solution." Cops had their own names for the stuff: "dirty dirt," "stinker juice," "Mr. Sousa," or just "Fred," as in "Fred B. Dead." Saturated with the noxious fluids of decay, the soil within a cadaver's drip zone reeks with an overpowering aroma unmatched in nature. Think skunk meets Montezuma's revenge. It's the last thing a crime technician would want to bring back to the laboratory. Even the disgust-power of squirming maggots pales in comparison. But what if "dirty dirt" could tell you when a victim died? Not just the year, month, or week. What if it could tell you the day of murder?

On rare occasion, forensic entomology could be that precise. Yet there remained countless situations where barriers such as closed windows and cold weather delayed or prevented insects from ever reaching a corpse. By contrast, the soil-solution work of Arpad Vass relied on two commodities always present at the moment of death—the corpse and its gut full of bacteria.

In the summer of 1988, the slight, young lab technologist had taken a break from his forensic science studies at the Medical College of Virginia to spend a morning among the bubbling casseroles and stainless steel dissecting trays of Bass's eerie laboratory beneath the stands of Neyland Stadium. By his own admission, Vass wasn't entirely sure why he had come. He'd never taken so much as a first-year anthropology class. Yet for months he'd been burning to talk to Bass about the professor's infamous 113-year error in the disinterment of Colonel Shy.

Bass had never thought to hide what a lesser anthropologist might have considered a humiliating error—that of mistaking a Civil War corpse for a modern murder victim. Then again, he had never dreamed the story would continue to circle the globe in both the popular and scientific press. A decade after the fact, Vass had read of it in a local paper, and it had started him thinking. Did the professor realize that the toxic metals in the Civil War hero's solid lead coffin could have been the sole reason for the body's strangely preserved state? "The lead would have effectively sterilized the body by poisoning the bacteria and other micro-organisms responsible for putrefaction," Vass explained. "So decomposition would have failed to progress past initial autolysis." In other words, the colonel might be dead meat, but without bacteria, he would never spoil. Not even embalming could stave off bacterial putrefaction for over a hundred years, Vass continued. For that, you needed a truly long-lasting poison—like the toxic metal continuously leaching out of Shy's solid lead coffin. Preservation by lead poisoning, as it were. The professor appreciated the insight, and the two men promised to stay in touch.

In the months that followed, the young microbiologist found his thoughts continually returning to death and decay. Admittedly, the monotony of his laboratory work left ample room for a wandering mind. Vass craved a new challenge that could make better use of his string of biology, medical technology, and forensic science degrees. He began scanning the scientific literature for time-of-death research, thinking that this might be an area where he could make his mark. The paucity of studies confirmed his impression, and further inspired his thoughts.

If insects overwhelm a corpse in identifiable waves, he reasoned, why not microbes? The succession might start with the aerobic, or oxygen-dependent, organisms already residing on skin and mucus membranes. Then, in the hours after death, the anaerobic bacteria of the human bowel would begin their staggering population explosion. After years of peacefully biding their time, they would rapidly spread through the stagnant bloodstream and the expanding ocean of dead and ruptured cells. Once bacterial putrefaction had finished melting away soft tissue, soil bacteria and wind-borne fungi might move in to colonize and crumble the dry remains.

"You might be onto something," Professor Bass agreed when the two men next spoke. Clearly, the UT Anthropological Research Facility would be the place to find out. What Bass did not have, however, was a paid position in his graduate program. Dozens of the nation's brightest anthropology students had already duked it out for the year's prize slots.

Only a caffeine-driven workaholic could appreciate the solution that presented itself. Vass snagged a full-time job in the University of Tennessee Microbiology Department, moved with his wife to Knoxville, and simultaneously launched himself into full-time studies as an unpaid doctoral student in the Department of Anthropology. Bass proved more than supportive, promising Vass a steady stream of research subjects and his own "patch" at the Body Farm.

Vass needed to know the *exact* time of death for every corpse, to the hour, not just the day. Not a problem, his new professor assured. In recent years, Bass's ragtag collection of unclaimed homicide victims and derelicts had been supplemented by a reliable stream of hospital patients who had bequeathed their bodies to science, some of them specifically to the Body Farm's provocative research. Every subject would come with a time-stamped death certificate.

He'd likewise need virgin ground, Vass persisted, lest cross-contamination from a previous corpse skew his results. Bass came up with that too—a large, gently sloping area north of the wire cages where Rodriguez had conducted his insect collections nearly a decade before. The plot would prove to be an ideal research site as well as a rejuvenating idyll after eight hours spent shuffling test tubes in the UT microbiology lab. Its wonderful mix of woodsy microhabitats included bare ground, honeysuckle-tangled brush, a few fallen trees, and a sun-dappled area of mayflowers and wild scallions.

There, on the first day of spring 1988, Vass began his studies with the assistance of a sixtyish white male who had died of heart failure the previous morning. His initial fear—that microbes might prove too scant in the very early stages of decay—proved amazingly naive. Vass found himself inundated. From the skin surface alone, he isolated dozens upon dozens of species of *Staphylococcus, Streptococcus, Candida, Malassezia,* and *Bacillus,* many of them no doubt present long before death. As he'd expected, the numbers and variety of organisms mushroomed with the onset of putrefaction, shifting to a preponderance of anaerobic bacteria—over a hundred species, including infamously toxic strains of *Salmonella, Serratia, Shigella, and Klebsiella,* as well as the ubiquitous gut fauna *Enterobacter and Escherichia coli,* the free-swimming *Proteus,* the gliding *Cytophaga,* and several species of amorphous amoebae.

Over the second and third week, the cadaver bloomed with countless other microbes carried by the feet and mouthparts of flies, ants, and beetles. The fauna of death filled Vass's petri dishes with their colorful efflorescence and crowded his incubators to overflowing. With everything from *Agrobacterium* to *Zooglea* showing up beneath his microscope, Vass began to wonder if there was any microbe that *didn't* play a role in human decomposition.

It's not working, a sleep-deprived Vass confessed to his professor at the end of week 3, or rather, it's working too well. Simply identifying the microbes was beginning to look like a lifetime undertaking. Even then, what hope would there be to forge a straightforward protocol for a typical crime lab? For that was Vass's goal: to provide forensic scientists and technicians with a simple, reliable method of determining time since death with standard equipment and minimal fuss. But with literally hundreds of bacteria, fungi, and other microbes to culture and isolate, by the time a lab technician had them all identified and correlated, the perpetrator could himself be dead. Vass wanted to start over from scratch.

If charting the fauna of death proved too complicated, he reasoned, perhaps their by-products would prove more amenable to study. The *molecules* of death, as it were, might likewise appear and vanish again at a steady, predictable rate. After all, the bacterial putrefaction of the human body was nothing more than the progressive breakdown of complex biological compounds into ever smaller and simpler ones. Proteins break down into amino acids, fats and amino acids into fatty acids, fatty acids into phenols and glycerol, and so on, till one reached the level of carbon dioxide and water. Just as the body builds itself out of these simplest of chemical building blocks, decomposition breaks it back down again. Deevolution. Dust to dust.

Bass liked the new direction. Another of his students had been attempting to chart something remotely similar, the postmortem decay rate of DNA molecules in rib bones. But that research remained bogged down in complicated laboratory procedures that prevented its testing in the field. Indeed, the only comparable method ever put into forensic use could be traced to the late 1960s, when British pathologist Bernard Knight developed a battery of a half-dozen tests for dating skeletal remains based on their gradual demineralization and loss of fatty marrow. In a test of sixty-eight bone samples, with known ages between 1 and 3,000 years, Knight found he could group them into broad groups of "less than 5 years," "less than 50 years," "less than 100 years," "less than 350 years," and "above." It was the sort of exhaustive testing that Bass could have used to distinguish Shy's Civil War bones from those of a modern homicide. In fact, in the early 1970s Knight had assisted forensic anthropologist Clyde Snow, one of Bass's rare peers, in one of his many investigations of human rights abuses and mass murder.

Specifically, Snow had been attempting to verify allegations that prison guards at an Arkansas state penitentiary were killing inmates on a regular basis and stashing their bodies in a nineteenth-century slave cemetery behind the prison. The story of the alleged murders and that of the crusading prison warden who fought to end Arkansas's legacy of prison brutality would inspire the film *Bruebaker.* Knight did in fact confirm that some of the skeletonized remains were of contemporary origin. Yet his bone dating methods, with their error margin of "give or take 50 years," could do no more than that—resolve whether a bone was a matter for the forensic scientist or the archaeologist.

By contrast, a chemical yardstick of microbial by-products would, in theory, begin ticking as soon as death threw open the gates to bacterial growth. Success would hinge on whether Vass could find a handful of molecules at levels that vary systematically over the days, weeks, and months since death. Ideally, the amount of each chemical would change at a different rate than the others, so that each day after death would correlate with a unique chemical pattern.

Vass's first challenge lay in understanding enough about decomposition to pick out which chemicals would prove informative and then finding or designing the appropriate tests to assay their levels. Textbooks could provide the raw basics: Within minutes of death, the cells of the human body begin to split and burst, disgorging their nutrient-rich contents of proteins, lipids, and carbohydrates. The first to the feast are the bacteria of the colon, or large intestine. They begin their growth by gorging on carbohydrates, which they ferment, or anaerobically digest, to yield sweet-smelling ketones, rotten-egg sulfides, whiskey-hued butanol, and copious amounts of methane. Importantly, Vass could separate and measure such chemicals easily with a standard laboratory chromatograph.

The microbial digestion of body proteins might prove even more informative, Vass reasoned. Bacteria must break down the long and complex protein molecule in several steps, each yielding a different set of by-products that would appear and then disappear as the breakdown continued. Before a bacterium can bring a protein inside its cell for digestion, it must exude enzymes to chop the long carbon backbone of the molecule into "bite-size" pieces. Once inside the bacterial cell, protein fragments meet digestive enzymes that continue their degradation. By slicing off a carboxyl group here and an amine or sulfide group there, digestion produces an array of simpler compounds, including fatty acids such as pyruvic and butyric acid and ptomaines such as the aptly named cadaverine and putrescine. The leftover sulfides, amines, and other chemical bits and pieces themselves react to produce mineral precipitates such as iron sulfide and nitrogen-rich ammonium.

The microbes consume body fats along a similar route, breaking down their branching long-chain molecules into shorter, simpler fatty acids and various alcohols, of which the sweet-smelling glycerol is the most abundant. Some bacteria can degrade the glycerol further, into acetone and phosphates; others render fatty acids into still simpler compounds such as vinegary acetic acid.

As these microbial reactions liquefy the body's soft tissues (putrefaction), their by-products purge from orifices and wounds, along with the spilled sodium and potassium salts that once conducted electricity through the living cells. Under the right conditions, the electrolyte salts react with the remaining fatty acids to produce adipocere, or grave wax, the pasty and crumbly soap that forms beneath and around a decaying corpse. This in turn becomes a banquet for soil bacteria and wind-borne fungi, poised to continue the biochemical breakdown. Bone follows its own complex pathway, with the degradation of its organic marrow and the leaching of bone minerals such as calcium and magnesium.

From this chemical zoo, Vass selected the molecules he hoped would prove the ideal time-of-death markers. For his studies of early decay—that is, the decomposition of soft tissue—he chose to chart a handful of volatile, or short-chain, fatty acids. By their chemical structure, Vass predicted they would remain stable enough to linger for days to weeks, yet eventually evaporate—hopefully at different rates. Moreover, the primary fatty acids of human decay—butyric, propionic and valeric acids—did not otherwise occur in the environment, unless a corpse happened to drop on top of an animal carcass or a bar of lye soap. Consequently, Vass would not have to worry about contamination from environmental sources. Just as important, any qualified crime technician could measure Vass's fatty-acid time markers with a gas chromatograph.

For his time markers of the extended postmortem period, or skeletonization, Vass chose a dozen minerals and electrolytes such as calcium and magnesium, and ions of potassium, sodium, chloride, and sulfate. Impervious to further microbial breakdown, they should linger in the environment around the corpse for weeks to months and possibly years. And they too could be detected and measured with a standard laboratory chromatograph.

Best of all, perhaps, Vass knew he had a perfect collecting medium—dirt. Ubiquitous and superbly absorbent, Mother Earth seemed custom-designed to sop up the by-products of human decay. Indeed, life on so-called dry land would prove impossible without soil's powerful hold on the nutrient-rich fluids suspended between its grains. It was this "soil solution" that Vass aimed to extract and test for his chemical time markers. Later, he could adapt his methods to sop up death's effluent from mattresses, carpeting, and the like.

Vass had his materials and methods assembled by the time another victim of heart failure came his way via the cooler of the medical center morgue. On a sweltering August afternoon, he slipped his chilly volunteer from its body bag onto a sunny patch of grass, trying hard not to appear awkward as his professor explained the purpose of the research to a cluster of visiting FBI agents. But what Vass saw next drop-kicked his heart into his throat. "My God, he's perspiring!" he yelped, staring dumbfounded at the prostrate corpse. As soon as his blood pressure dropped back under 300, the chagrined grad student realized his mistake. Dead men don't sweat. However, a corpse refrigerated at 48 degrees Fahrenheit will glisten with dew on a hot and humid Tennessee day.

For the most part, Vass's field studies proceeded peacefully thereafter, save for an afternoon spent chasing and being chased by a rat unwilling to share what was left of an emaciated cancer victim. In the shade of the farm's tulip poplars and mulberry trees, the sleep-deprived grad student processed his samples, sifting the greasy dirt through a steel sieve, weighing it with a battery-operated balance, preparing a standardized solution of soil and deionized water, and packing the mud into centrifuge tubes. Slipping his samples into an ice-filled picnic cooler, he lugged them back to campus each night to complete his analysis in the nighttime quiet of the microbiology lab.

Autumn rains brought good news. Even a torrential downpour didn't wash away death's chemical markers. Quite the opposite, the slimy by-products of putrefaction seemed to bind the soil together like a waxy, water-repellent glue. True to his word, Bass kept the test subjects coming at a steady rate. By the following spring, Vass had joined the small cadre of scientists intimately familiar with the seasonal cycles of human decay, from the blistering speed of summer to the slow mummification and molder of winter.

Vass noted that the progression of his soil markers likewise sped up and slowed down with the seasons. In summer, fatty acid levels rapidly peaked and plummeted again within a few weeks, whereas in winter their levels continued to slowly build for months on end. The minerals leaching from the cadaver into the soil followed a similar pattern. Clearly, their levels were keeping pace not with the calendar but with the stages of decay, which in turn accelerated and braked with rising and falling temperatures. What he needed was some sort of mathematical factor or regression formula to convert the chemical process back to chronological time.

The answer came during the second summer of research by way of Neal Haskell, who had come to set up shop on the hillside across from Vass's quiet idyll. "So you're the stinker juice guy!" Haskell guffawed in introduction. Bass had forewarned his grad student that he'd be getting company. But nothing could quite prepare him for the larger-than-life entomologist. Haskell's mission that summer was to establish, once and for all, the validity of using freshly slaughtered pigs as experimental stand-ins for homicide and so put an end to courtroom challenges of the method's applicability to murder. He would do so by comparing the arrival, growth, and succession of insects on side-by-side pig carcasses and human cadavers, the latter available nowhere else but Bass's four-acre wood.

The burly Midwesterner and the slight, Eastern European grad student made an odd couple sharing a pitcher of beer that weekend. Far stranger was their conversation, which Haskell delighted in broadcasting across a room for its inevitable shock effect. Though Vass found the splattery details of Haskell's murder cases interesting enough, his real interest lay in finding some way to mathematically tame the wild influence of temperature on his data.

"Hell, Arpad! What you've got here is a simple problem of accumulated degree days," the entomologist insisted. Sketching out the uncomplicated arithmetic on a bar napkin, Haskell explained the simple formula that he had long used to calculate maggot growth. "Ten November days at an average temperature of 10 degrees [Celsius] or two summer days at 50 degrees. It doesn't make a damn bit of difference. Both add up to 100 accumulated degree days to a maggot. Now why can't you do the same thing with the stuff in your stinker juice?"

Vass found the concept almost too simplistic. Still, he saw the clear parallel between the heat-fueled metabolism of Haskell's blow flies and that of the microbes churning out the decay products of his own research. Over the next week, he ran the numbers, factoring the soil concentrations of each fatty acid and decay mineral against daily local temperatures for the previous year. Like magic, the soil-solution levels for all six cadavers, in all four seasons, fell into sync.

When charted across a graph with accumulated degree days (ADD) as its base, the soil levels of each fatty acid traced its own distinct roller-coaster path between 0 and 1,200 ADD. At which point, they all crashed back to 0—the end of soft-tissue decomposition. For their part, the levels of inorganic soil markers began their rise too steeply and close together to be useful, but spread out around 700 ADD to pick up where the fatty-acid ratios left off and extend the postmortem time line past 3,000 ADD, where Vass's year-old study now stood. (They would continue to prove informative over a second year of study.) Consequently, any point on the time line corresponded to a unique ratio of chemical markers. Measure the molecules in the soil beneath any corpse, calculate the levels against daily temperatures, et voilà, you have days since death.

Over the following months, Vass tightened his time estimates still further with a mathematical standard to adjust for gross differences in predeath body weight, since different proportions of fat and muscle appeared to produce slightly different concentrations of volatile fatty acids.

By the fall, Bass pronounced it high time for his apprentice to begin working on actual murders. Specifically, he wanted Vass's help on several cases in which he didn't have a clue as to time since death. One of the first involved skeletonized remains found lying in a ditch. Bass had succeeded in identifying the victim, a teenage boy, through medical and dental records. He turned to his graduate student for time of death.

Vass prepared his soil solution from a fat plug of dirt pulled from the ground beneath the bones. With soft tissue a nonentity, he focused his gas chromatograph on the inorganic tracers. The sharp peaks on the paper readout ticked off 27.9 parts per million (ppm) of sulfate, 1.4 ppm chloride, 10.2 ppm sodium, 6.9 ppm potassium, and 7.1 ppm calcium. The absence of any appreciable amount of either ammonium or magnesium told Vass that the boy had been dead less than 3,250 accumulated degree days. The soil concentration of sulfate, adjusted for the boy's known weight of ninety-six pounds, factored out to approximately 3,000 accumulated degree days. Running the numbers on the remaining minerals, Vass ended up with a range of 2,250 and 3,000 ADD. Translated against weather records of daily temperatures, the boy had been dead between 168 and 183 days.

The two-week window told police the boy had been killed shortly after disappearing from home. More important, it took forensic science into uncharted territory. Given the evidence—a heap of disarticulated bones and a scattering of pupal cases from long-departed flies—the combined efforts of the world's best forensic pathologists, anthropologists, and entomologists could not have narrowed time of death to within two months, let alone two weeks.

Vass proved he could do even better when presented with a still-dripping corpse, like that of the unidentified Tennessee man he helped Bass scrape from a farm field on an otherwise exquisite spring morning. Once done with the messy work of "bagging," Vass returned to the greasy outline that marked where the body had lain. Into the "heart" of that dark stain, he drove a three-inch-wide aluminum pipe to extract a specimen for processing back at the lab.

Wetting, centrifuging, and analyzing his sample, Vass ran the numbers on five fatty acids, adjusting his results for what he judged to have been a slightly built man, approximately 100 to 150 pounds. Their levels all fell between 675 to 775 accumulated degree days, giving Vass a postmortem interval of forty-one to forty-eight days when adjusted for daily temperatures. The seven-day interval was more than enough to match the remains to a missing person report. The man had last been seen alive fifty-two days before the discovery of his remains.

Similar cases followed as Vass wrapped up his graduate work at the University of Tennessee. Although many cases proved invaluable in guiding police investigations, none made its way to the testing grounds of the American courtroom, where opposing lawyers make or break new forensic science in pretrial admissibility hearings. Only when Vass's methods had overcome that hurdle—designed, in essence, to discredit them—could they become part of the accepted procedures of homicide investigation.

Meanwhile, the newly graduated Dr. Vass needed to find a real job. That is, one paying enough to support a family that, with baby, now numbered three. Fortunately, his new position in the Life Sciences Division of Tennessee's Oak Ridge National Laboratory didn't take him far from what he considered his life's calling. Nor did his supervisors begrudge him the occasional afternoon off from his work studying the microbial contaminants of drinking fountains and eyewash stations. All they asked was that he keep the ventilation fans running on high while processing the wretched stuff he brought back from his field trips.

In December 1991, the *Journal of Forensic Science* accepted Vass's time-of-death research for publication, with William Bass in the professor's revered place as second author. Still, Vass continued to narrow his postmortem estimates with more sophisticated mathematics and added parameters such as soil pH and particle size. When Thomas "Zoo Man" Huskey made the *Knoxville Nightly News* in October 1992, Vass had his postmortem determinations down to plus or minus a day for every week past death. Bass called him onto the case the night that police discovered a fourth body.

The next morning, Vass packed his ice chest and collecting jars, left a message with his supervisor that he would be late for work, and set off for a morning hike in the woods behind Cahaba Lane.

Bass and the Knoxville medical examiner had already divvied up the victims: two to the morgue and a second pair—too far gone for autopsy—loaded into the professor's truck for transfer to the Body Farm.

Thanks to the tenacious fluids of decay, Vass did not need a chalk outline to see where each young woman had lain. He did, however, need to work quickly, before the day's windfall of autumn leaves covered their greasy stains. He spaded the disturbed soil of the shallow grave behind the billboard into a glass jar, stashed it on ice, then followed a yellow ribbon of police tape into the woods, where just over the rise of a small hill he found a second broad smudge of cheesy residue. The second jar filled and iced, he moved south to the top of a small ravine, where the splintered stump of a small tree marked where a third victim had been tied. At some point, the trunk had snapped, tumbling both tree and corpse into a small creek at the bottom of the gully. There Vass could see yet another telltale stain, extending across the muddy streambed.

This was something new. Would the groundwater skew his readings? Rain hadn't done so during his field studies at the Body Farm. Vass filled his third jar with the greasy mud with high hopes of what might emerge from his laboratory centrifuge. A bright circle of police tape on a distant tree marked his final destination, the spot where the police had found the fourth victim tied and slashed to death. Vass made his final collection from the soil between the roots of the tree.

Before leaving, Vass verified that police would be taking the onsite temperature readings he would need to adjust data from the nearest weather station. Haskell would need the same information when he arrived from Indiana that evening. Such a visit would normally be grounds for a round of beer. But for now, the two men would deliberately avoid any communication. Not until each had handed over his results to Bass would they compare notes.

As it turned out, their dates dovetailed precisely, with Vass's narrower time ranges falling dead center within Haskell's broader estimates. Vass pinpointed the death of the first victim, the corpse found sprawled across the creek, to September 25, give or take three days. Much to his delight, the sogginess of her deathbed had not made an appreciable difference in results. The second and third killings, found nearest the road, he pinned to October 13 and 15, give or take two days, and that of the last victim, found tied to the tree far back in the woods, to October 20, give or take one day.

This could be *the case,* Haskell predicted, the case that could establish a legal precedent for his young friend's methods and make them the subject of conversation in every crime lab across the country. It did not hurt that the case was becoming national tabloid material. Adding another bizarre twist to the "Zoo Man" murders, Thomas Huskey's lawyers were claiming their client suffered from dissociative identity, or "multiple personality," disorder. His initial confession to the murders, they claimed, were in fact made by his evil "alter ego" Kyle, who'd also planted evidence to frame "Tom" for the crimes.

Moreover, time of death promised to play a prominent role in the trial to come. The defense had quickly seized on the medical examiner's initial statement that the last body discovered, on October 26, appeared to be just twenty-four to forty-eight hours old. If that were the case, Huskey would have already been in custody for the first three murders. By contrast, Vass and Haskell had independently calculated the death as occurring on the day or two immediately before the arrest.

It would be over six years before Huskey finally came to trial. In the end, a skittish district attorney decided against introducing Vass's "futuristic" soil chemistry methods in favor of Haskell's tried-and-true entomology. It seemed a prudent enough move, given the virtual guarantee that Vass's new techniques would trigger a drawn-out legal challenge from the defense. Though

Haskell's bugs could not, by nature, be as precise as Vass's microbial chemistry, they would support the prosecution's claim that the final victim, Patricia Johnson, had been killed the day before Huskey's arrest.

What the prosecution did not count on was the defense coming up with its own entomologist to challenge Haskell's results. Flown in from Hawaii for the trial, Lee Goff claimed that Haskell's work was off by at least three days. Given forensic entomology's margin of error, their quibbling appeared to cast reasonable doubt as to whether Huskey could have committed the final murder. The trial ended in a hung jury, and Huskey remains in a state mental hospital awaiting retrial. Vass and his time-of-death determinations stand ready for subpoena.

12 THE NEW MOD SQUAD

Believe those who are seeking the truth.
Doubt those who find it.
—ANDRÉ GIDE (1869–1951)

STEPPING INTO THE early March twilight, Jason Byrd drank in the silence, grateful to escape the drone of a thousand flies. For the most part, the twenty-six-year-old graduate student thought of his laboratory at the University of Florida as a haven, a refuge where hours disappeared like minutes as he worked through his forensic cases and tended his insect colonies. Most visitors wrinkled their noses at the smell—a cinnamony bouquet of fermenting sugar water, dead bugs, and raw pork past its prime. Mercifully, the hairlike olfactory nerves of the human nose tire easily. Habituated to the odor, Byrd smelled nothing within minutes of unlocking his door each morning.

The same could not be said of the incessant buzz that radiated from the racks of wire-mesh rearing cages that held Byrd's research colonies. Compounding the abrasive noise of blow fly wings was the constant whirr and suck of the fume hoods that vented the lab's sickly sweet air before it could permeate the entire entomology building. The jangle of the laboratory telephone had felt welcome that afternoon, like a pure note breaking through cacophony. Koren Colbert, crime-scene technician for the Bay County sheriff's office, was calling from Panama City. She'd be driving down to the Gainesville campus by nightfall to deliver the contents of a grave discovered in the piney woods that rise up behind the Florida panhandle's sugar-sand beaches.

Three hours later, Byrd gladly took a break from his laboratory's maddening buzz to meet the ponytailed crime technician in the campus parking lot, where they unloaded her evidence van. In the morning, Byrd promised her, he would drive the largest of the three cartons across town to the office of forensic botanist David Hall, before checking in with forensic anthropologist William Maples at his UF laboratory, a stone's throw away from the entomology building.

For it wasn't just insects that Colbert was delivering that evening. Byrd was the point man of sorts for a veritable Mod Squad of biologists who would tackle the case together from the triple perspectives of entomology, botany, and anthropology. By most measures Byrd was the junior member of the team, but it was his vision that had catalyzed their multidisciplinary approach five years earlier.

As a freshman in 1990, Byrd had assembled for himself an unprecedented faculty advisory committee. He had come to the University of Florida to fulfill his dream of combining a farm boy's appreciation of nature's recycling efforts with a burning desire to follow in the footsteps of the rural sheriffs who had been his childhood role models. He aimed to become a death investigator the likes of which the world had never known—able to appreciate the significance of every root, tendril, bug, and bone found at the scene of murder.

Through eight years of study, Byrd honed a special expertise in death's insect fauna. In doing so, he stood out among the second-generation of forensic entomologists who were advancing their science beyond the foundation laid in the 1980s. The field's early pioneers had bent their understanding of agricultural pests and disease-carrying insects to the service of estimating time of death. Forensic entomology's second generation could now go further, by studying necrophilous insects with the sole purpose of forging a more accurate and far-reaching postmortem clock.

Ironically, perhaps, it was by poking holes in the work of their predecessors that Byrd's generation would advance their science. Huge gaps still existed in what was known about the basic biology of many forensically important insects. Moreover, computer models of larval growth and adult behavior—such as that now being developed by Byrd as his graduate project—threw new light on the many variables that could speed up or slow down any living timepiece.

Early on, Byrd had seen the advantage, even the necessity, of viewing the human corpse as more than insect fodder. A complex and dynamic ecosystem, it simultaneously encompassed the chemical breakdown of life and its subsequent rebirth. Entomologists and botanists could both time the latter—that is, the new life that rose from and around this fleeting natural resource. From the opposite perspective, the pathologist and anthropologist could shed light on the progressive melting away of flesh and crumbling of bone. No one of these sciences could reliably predict time of death in all, or even most, circumstances. Yet taken together, their individual contributions could weave a telling tale.

It was with this perspective that Byrd appeared in Hall's mail-cluttered office the morning after Colbert's delivery, wrangling an unwieldy, three-foot-high carton through the doorway. Setting down the box and splitting the red band of evidence tape, Byrd hoped he might finally have something to stump his faculty adviser, if only for a few minutes. He related the known facts as he pulled the spindly tree out of its container: Colbert had noticed the sapling growing up through a bundle of ropes a few feet away from the clandestine grave. In particular, she noted that the ropes in the bundle matched those tied around the ankle bones and wrists of the skeletonized remains. All had been hand-tipped with the same kind of blue, heat-shrink material.

Hall grabbed the sapling with one hand as he raised his magnifying loupe to his eye with the other. "Yaupon holly, *Ilex vomitoria.*"

"So much for providing you with a challenge," Byrd shrugged. From the bottom of the box, he retrieved several zip-lock bags, the contents of which Hall promptly identified as deer moss and newly germinated persimmon seeds. The labels on the bags specified that the plant material had come from the soil directly over the grave. Consequently, that grave had to have been closed for a minumum of three weeks, Hall concluded. But that much was already clear from the bareness of the bones that Colbert had pulled from the grave. The skeletonized remains, already identified by Maples as those of a young woman, now lay on the anthropologist's examining table, waiting for Hall and Byrd to complete their estimate of postmortem interval. With that determination in hand, the anthropologist could begin combing the medical records of persons reported missing around the same time—looking for the right set of measurements and prior injuries to make a positive identification.

The yaupon holly might provide the year. For the plant to work its way up between the individual ropes, it must have been a mere seedling when the bundle landed on top of it, Hall reasoned. At nearly three feet, the tree had more than enough time to establish itself. The question was, How much time? Hall knew the species to be an amazingly fast grower.

Grabbing a pair of pruners, the botanist made a clean cut through the sapling's trunk four inches above its ball of roots. He showed Byrd a thin pale band of light wood just under the bark. It represented the rapid spring growth of the previous month, he explained. Then came a dark ring denoting the past winter, followed by a much fatter circle of light-colored wood—a complete growing season the previous year, 1994. At the center, a pithy disk represented the holly's first flush of growth—beginning sometime in

1993. Hall assigned Byrd the task of checking weather records for the two years in question for any periods of drought or unseasonable weather that might have produced a false ring or obliterated an expected one.

Back at his campus laboratory, Byrd found a three-inch-thick weather service report waiting on his desk. Nothing in the rundown of daily temperatures and precipitation appeared severe enough to change Hall's estimate. So Byrd called Colbert with the year of death, promising to update her by the end of the day if the bugs narrowed the time window further.

Byrd quickly identified the mass of empty pupal cases collected from the grave as those of the black blow fly, *Phormia regina,* a chunky, greenish-black bruiser that dipped as far south as Florida only in the coolest months of the year. Temperatures over 80 degrees sent it skittering northward, whereas anything much lower than 50 degrees grounded the fleet. By themselves, the empty pupal cases told Byrd no more than that the postmortem interval had to have been long enough for the maggots to complete their ten- to twenty-day life cycle. But Hall's bracketing of the year gave the entomologist a specific time period to examine for fly-friendly weather.

He began to work backward through the daily weather reports beginning in the spring of 1994, by which time, Hall had determined, the yaupon holly had snaked its way up through the telltale ropes and laid down its first annual ring. Byrd saw that the winter of 1993–1994 had remained uncommonly harsh, beginning with a hard freeze in early November and remaining too cold for fly activity through March 1994. Therefore, Byrd began working the time window from the other end.

Early in the spring of 1993, temperatures climbed into the 80s—too hot for black blow flies—and remained there through the end of September. The second week in October brought the first break in the sweltering heat, with daily temperatures ranging from the upper 40s to the lower 80s. Black blow fly heaven. Their numbers would have climbed to teeming levels before an early November cold snap wiped the slate clean again. Byrd called Colbert with his two-week estimate, mid- to late October 1993.

The Bay County investigator took the news with a leaden sense of finality. From Maples, she had just learned that the bones spoke of a young woman with multiple fractures fully healed before death. Colbert had thought immediately of a local prostitute who'd been beaten savagely several years earlier. The same prostitute had disappeared in the fall of 1993. Her parents had clung to the hope that their daughter had left to start a better life. Now it was Colbert's hateful task to show them the rings she had forced herself to remove from the grave's bony fingers. Maples would make the final identification from childhood dental X rays.

In the late 1990s, Byrd and Hall would organize their unique collaborative approach into an annual death-investigation workshop for cops and coroners. "Bugs, Bones, and Botany" they called it. Various forensic anthropologists rounded out the teaching crew after Maples's death in 1997. In essence, their aim has been to teach their frontline colleagues to process any death scene or corpse as if it were an ecological field study. That is, with an eye for every living and once-living clue of the time elapsed since death.

At the beginning of the twenty-first century, progress in each of their three interrelated fields depends on such at-the-scene thoroughness. The usefulness of Byrd's computer model of insect development, for example, hinges on his receiving a representative sampling of every kind and size of maggot found on or near the body. No matter how sophisticated his programming of mathematical regression formulas, a skewed collection will produce skewed results.

No less important, such workshops serve as a kind of outreach effort to law enforcement that has led to new money for continued research. In 2001, for instance, the National Institutes of Justice agreed to fund phase 2 of Byrd's bugs-on-bodies computer model. The revised model will encorporate the effects of predaceous insects such as fire ants and hairy maggots, which can reset the postmortem clock by clearing a corpse of previously laid eggs and less-aggressive fly larvae. The ultimate aim is a web-based model that trained users the world over can access to estimate postmortem interval from their own cadaver insect collections.

The justice institute is likewise funding research by entomologist Jeffrey Wells of the University of Alabama, who, like Germany's Mark Benecke and Canada's Felix Sperling, is developing ways to identify fly eggs and early instar maggots with DNA amplification and typing. Wells's federally funded work has already produced DNA profiles for thirty-five forensically important flies, many of which have remained virtually impossible to distinguish by appearance alone, especially in their earliest stages. Despite their outward similarities, such closely related species can have dramatically different life cycles—arriving at a corpse at different times and maturing at different rates. As a result, their genetic identification has the potential to dramatically improve forensic entomology's contribution to death investigation. At the same time, with DNA typing kits now costing as little as $30 to $40, the work of Wells, Benecke, and Sperling may soon enable forensic labs to perform their own entomological death estimates. "You don't have to be an entomologist to grind up a bug and generate a DNA profile," Wells notes.

Genetic typing stands poised to likewise revolutionize estimates of time since death based on plant growth. "Lots of times I'm handed a shank of roots pulled from a grave, with no hopes in Hades to figure out the species," explains David Hall. Even two outwardly identical varieties of a hybridized plant can have dramatically different growth rates that would affect the botanist's estimate of time since death. The technology stands ready; millions of dollars have already been invested in the DNA typing of patented crops by companies needing a definitive test to distinguish their seeds from less valuable but outwardly identical varieties. The need remains for someone to generate DNA profiles for the more prosaic plants likely to show up in a death investigation.

Meanwhile, Hall is preparing a grant proposal for the National Institute of Justice to replicate entomology-style pig studies from a botanical perspective. Rather than study the insect fauna erupting atop the mock-homicide victims, Hall aims to study the effect on underlying vegetation. "Once that corpse drops, we need to clock how long it takes for, say, a patch of grass to turn yellow," he explains. "How long before the grass dies completely? How long till it rots?" Similarly, much remains to be learned about the effect of a buried corpse on surrounding and overlying vegetation.

"Most people look at the flush of growth atop a grave and figure it's coming from the nutrients of the decomposing body," Hall notes. "In truth, the growth occurs simply because the soil has been disturbed. The fluids of decomposition, by contrast, can be quite caustic for some time after death." What no one knows, as yet, is just when these caustic decay fluids begin to stunt or otherwise affect surrounding plant growth and how long the effect lasts. The answers could provide still more markers for the postmortem time line.

Perhaps the most exciting progress in the age-old quest to pinpoint time of death is now coming out of Oak Ridge National Laboratory, which, thanks to Arpad Vass, has forged a close association with the forensic anthropology program at nearby University of Tennessee. Once Vass's Oak Ridge supervisors got over the unsettling odors emanating from the UT graduate's forensic sideline, they realized the huge role he could play in their retooling of the national lab's post–Cold War mission. As part of the sword-to-plowshare shift, Oak Ridge has begun building one of the most successful forensic science programs in the nation. Vass's time-of-death work promises to be a crowning jewel, with a grant from the Department of Energy to study new biochemical markers of postmortem interval. Vass is searching especially for highly accurate biomarkers for the first two weeks after death. As such, they would span the gap that now exists between pathology's triple timepieces of algor, livor, and rigor mortis, good only for the first twenty-four to forty-eight hours after death, and Vass's previously developed tests for cadaver leachate in the soil, which give the best results once a body has begun to liquefy. Moreover, Vass is gunning to get as close as possible to *hourly* markers for the time period in question, something never accomplished in the 2,000-year-old struggle to pinpoint time of death.

Assisting Vass at the Body Farm, UT graduate students Jennifer Love and Jennifer Synstelien have forged two very different sampling methods. Love's work involves a textbook-size box held over a corpse or possible grave to collect its gaseous emanations. The device holds a cluster of tiny balls the size of peppercorns. Made of absorbent aluminum-silica mesh, the balls act as a molecular sieve to collect decay gases that Love later decants into an aroma scanner, or electronic nose, in Vass's Oak Ridge laboratory. The "nose" does not identify the specific chemicals present in the gaseous mix. Rather, thirty-two chemical sensors within the scanner selectively react with one or more of ten different classes of chemicals such as amines, alcohols, ketones, ethers, and chlorinated hydrocarbons, to produce a distinctive "aroma pattern" of decay. Love had hoped to see the pattern change systematically over the hours and days after death. That she has not may simply mean that the work demands a more sensitive aroma scanner. "At a minimum," she says, "we've now shown that we can collect these odors and get them into the laboratory. We also see that their overall intensity changes over time."

More promising in the short term may be the work of the program's other Jennifer. Under Vass's direction, Synstelien has been camping out at the Body Farm to snip tissue samples of various organs and muscles at hourly and daily intervals over the first weeks after death. She and Vass then sort through the chemical components with a gas chromatograph and mass spectrometer to look for those that change systematically over time. After two years of research with more than 100 potential biomarkers from a half-dozen different organs and tissues, they have narrowed their search to the most promising candidates.

They are now closely monitoring the appearance and breakdown of twenty amino acids, using statistical analysis to find informative patterns as their ratios evolve over time since death. Surprisingly, the amounts of certain amino acids such as tryptophan seem to vary dramatically from one corpse to another. Other amino acids degrade too quickly to be of use as postmortem time markers. Most encouraging, Vass has noted a particularly stable relationship between the amino acids alanine and isoleucine, their ratio changing neatly over time, or more specifically over accumulated degree days (heat compounded by time). Biochemically, Vass cannot say what's going on to drive their changing levels. What's important, he says, is that their ratio to one another appears to shift at a predictable rate, at least in certain organs such as the heart and deep muscle. Vass sees other potential chemical clocks in the gradual breakdown of the amino acids lysine and ornithine into the odiferous decay chemicals putrescine and cadaverine, and that of the amino acid gamma-aminobutyric acid (GABA) into gamma-hydroxybutyrate (GHB), the latter infamous as the active ingredient in "date-rape" drugs.

Ideally, Vass would like to identify two relatively long-lasting organs or tissue areas as standardized sampling sites for such chemical time markers, so that injury to one leaves the crime technician or medical examiner with a second source. Like his earlier soil-solution test, Vass's ultimate goal is a straightforward test of postmortem interval that can be performed by any lab technician.

Meanwhile, pathologists the world over continue to search for more accurate measures of time for the very early postmortem interval that remains their bailiwick. Pathologists at the University of Glasgow have improved the measurement of algor mortis, or postmortem body cooling, with a system of multiprobe electronic thermometers thrust deep inside brain, liver, and rectum and monitored continuously with computer programs that compare the temperature readings with standard cooling curves derived from experiments with hundreds of covered and naked cadavers. But it remains to be seen whether medical examiners will invest in such equipment, given the modest improvement it affords in estimating postmortem interval.

Similarly, pathologists at Japan's Tottori University have developed a light-absorption meter that provides a more objective and potentially accurate measure of lividity, the gradual discoloration of the body caused by the gravitational settling and chemical breakdown of blood.

Researchers in at least a half-dozen countries continue to study postmortem eye chemistry, hoping to rehabilitate the once-bright prospects of correlating time since death with changes in the eye's vitreous humor. Some scientists say they are certain that the best prospects lie in improving the mathematical regression formulas used to calculate the eye's postmortem potassium levels. Others are turning away from vitreous potassium as the indicator of choice, to search for other decay products within the eye whose changing levels may prove more informative, or at least more consistent from one cadaver to the next.

Still others are pioneering entirely new tests for postmortem interval based on such things as the breakdown of nuclear and mitochondrial DNA and the declining viability of skin grafts removed at autopsy. Among the most promising is a test of bioelectrical impedence across various parts of a dead body. In essence, such a test measures the speed with which a small electric current passes through tissue. Because fat conducts electricity more easily than lean muscle, such tests have long been used on athletes, dieters, and other live subjects to calculate lean body mass. What pathologists noticed was that in the first twelve to twenty-four hours after death, the natural impedence of any tissue gradually increases, for the simple reason that electricity passes more slowly through colder tissues. After twenty-four hours, impedence drops again as dying cells spill their electrolyte-rich fluid contents, which help carry, or speed, the passage of electrical current. By measuring this deceleration and subsequent acceleration of current through tissue, postmortem tests of impedence become a continuous measure of body cooling (algor mortis) followed by cell destruction (autolysis) that continues for up to seventy-two hours after death.

Although each of these experimental measures of postmortem interval have passionate advocates, none have produced the kind of consistent and informative results that would warrant their widespread use. The world's leading forensic pathologists continue to urge caution and the liberal use of the word "unknown" on any form specifying the time of an unwitnessed death.

In the end, perhaps, any effort to medically *diagnose* time of death may be doomed to failure. For as any biologist knows, the demise of an organism as complex as a human—composed of some 100 trillion, semiautonomous cells—is anything but instantaneous, or even definitive. No one can explain why a puff of air and a thump on the chest brings one corpse back to life, whereas another—just as fresh and in similar condition—fails to respond to the most Herculean medical resuscitation efforts. The variation between any two human bodies only increases as we watch them disintegrate in the days, weeks, and months after the last heartbeat.

Far more promise may lie in the newer efforts to grasp the ecology of death. When the German naturalist Ernst Haeckel coined the term *ecology* in 1869, he described it as the wholistic study of living systems interacting with their environment. Ecologists look at communities of organisms, patterns of life, natural cycles, and population changes. And that is precisely what a new generation of forensic entomologists, botanists, and anthropologists have begun to do: They step back and appreciate the human corpse as a natural resource, even an ecosystem, teeming with life and responsive to its physical environment. From this perspective, death has many clocks.

Further Reading

For those interested in reading more about the sciences behind time-of-death determinations, the author recommends the following academic and professional references:

Spitz and Fisher's Medicolegal Investigation of Death, 3d ed., edited by Werner Spitz. Springfield, Ill.: Charles C. Thomas, 1993.

The Estimation of the Time Since Death in the Early Postmortem Period, by Claus Henssge, Bernard Knight, Thomas Krompecher, Burkhard Madea, and Leonard Nokes. London, Boston, Melbourne, Auckland: Edward Arnold, 1995.

Time of Death, Decomposition, and Identification: An Atlas, by Jay Dix and Michael Graham. Boca Raton, Fla., London, New York, Washington, D.C.: CRC Press, 2000.

Forensic Taphonomy (the anthropological study of death assemblages), edited by William Haglund and Marcella Sorg. Boca Raton, Fla., London, New York, Washington, D.C.: CRC Press, 1996.

Forensic Entomology: The Utility of Arthropods in Legal Investigations, edited by Jason Byrd and James Castner. Boca Raton, Fla., London, New York, Washington, D.C.: CRC Press, 2000.

Entomology & Death: A Procedural Guide, edited by Paul Catts and Neal Haskell. Clemson, S.C.: Joyce's Print Shop Inc., 1990

A Manual of Forensic Entomology, by Kenneth Smith. Trustees of the British Museum, 1986.

Also read . . . for the pure delight of it:

Death to Dust: What Happens to Dead Bodies, 2d ed., by Kenneth Iserson. Tucson, Ariz.: Galen Press, 2001.

Index